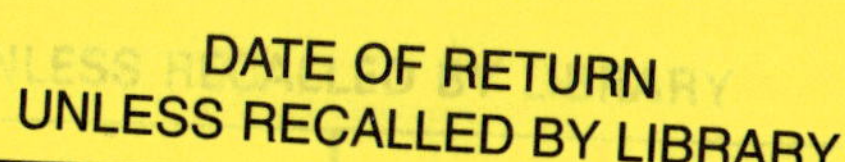

200137240

CABLE NETS AND TENSEGRIC SHELLS
Analysis and Design Applications

ELLIS HORWOOD SERIES IN CIVIL ENGINEERING

Series Editors
Structures: Professor H. R. EVANS, Department of Civil Engineering, University College, Cardiff
Hydraulic Engineering and Hydrology: Dr R. H. J. SELLIN, Department of Civil Engineering, University of Bristol
Geotechnics: Professor D. MUIR WOOD, Department of Civil Engineering, University of Glasgow

Addis, W.	**Structural Engineering: The Nature of Theory and Design**
Allwood, R.	**Techniques and Applications of Expert Systems in the Construction Industry**
Bhatt, P.	**Programming the Matrix Analysis of Skeletal Structures**
Blockley, D.I.	**The Nature of Structural Design and Safety**
Britto, A.M. & Gunn, M.J.	**Critical State Soil Mechanics via Finite Elements**
Bljuger, F.	**Design of Precast Concrete Structures**
Carmichael, D.G.	**Structural Modelling and Optimization**
Carmichael, D.G.	**Construction Engineering Networks**
Carmichael, D.G.	**Engineering Queues in Construction and Mining**
Cyras, A.A.	**Mathematical Models for the Analysis and Optimisation of Elastoplastic Systems**
Dowling, A.P. & Ffowcs-Williams, J.E.	**Sound and Sources of Sound**
Edwards, A.D. & Baker, G.	**Prestressed Concrete**
Fox, J.	**Transient Flow in Pipes, Open Channels and Sewers**
Graves-Smith, T.R.	**Linear Analysis of Frameworks**
Gronow, J.R., Schofield, A.N. & Jain, R.K.	**Land Disposal of Hazardous Waste**
Hendry, A.W., Sinha, B.A. & Davies, S.R.	**Loadbearing Brickwork Design, 2nd Edition**
Heyman, J.	**The Masonry Arch**
Holmes, M. & Martin, L. H.	**Analysis and Design of Structural Connections: Reinforced Concrete and Steel**
Iyengar, N.G.R.	**Structural Stability of Columns and Plates**
Jordaan, I.J.	**Probability for Engineering Decisions: A Bayesian Approach**
Kwiecinski, M.	**Plastic Design of Reinforced Slab-beam Structures**
Lencastre, A.	**Handbook of Hydraulic Engineering**
Leroueil, S., Magnan, J.P. & Tavenas, F.	**Embankments on Soft Clays**
MacLeod, I.	**Analytical Modelling of Structural Systems**
Megaw, T.M. & Bartlett, J.	**Tunnels: Planning, Design, Construction**
Melchers, R.E.	**Structural Reliability Analysis and Prediction**
Mrazik, A., Skaloud, M. & Tochacek, M.	**Plastic Design of Steel Structures**
Pavlovic, M.	**Applied Structural Continuum Mechanics: Elasticity**
Pavlovic, M.	**Applied Structural Continuum Mechanics: Plates**
Pavlovic, M.	**Applied Structural Continuum Mechanics: Shells**
Shaw, E.M.	**Engineering Hydrology Techniques in Practice**
Spillers, W.R.	**Introduction to Structures**
Vilnay, O.	**Cable Nets and Tensegric Shells: Analysis and Design Applications**
Zloković, G.	**Group Theory and G-Vector Spaces in Structural Analysis: Vibration, Stability and Statics**

CABLE NETS AND TENSEGRIC SHELLS

Analysis and Design Applications

OREN VILNAY B.Sc., M.Sc., D.Sc.
Department of Civil Engineering
Technion — Israel Institute of Technology, Haifa, Israel

ELLIS HORWOOD
NEW YORK LONDON TORONTO SYDNEY TOKYO SINGAPORE

First published in 1990 by
ELLIS HORWOOD LIMITED
Market Cross House, Cooper Street,
Chichester, West Sussex, PO19 1EB, England

A division of
Simon & Schuster International Group
A Paramount Communications Company

Typeset in Times by Ellis Horwood Limited
Printed and bound in Great Britain
by Bookcraft (Bath) Limited, Midsomer Norton, Avon

British Library Cataloguing in Publication Data

Vilnay, Oren
Cable nets and tensegric shells.
1. Cable-suspended net structures
I. Title
624.1774
ISBN 0–13–115726–4

Library of Congress Cataloging-in-Publication Data

Vilnay, Oren, 1941–
Cable nets and tensegric shells: analysis and design applications / Oren Vilnay.
p. cm. — (Ellis Horwood series in civil engineering)
Includes bibliographical references and index.
ISBN 0–13–115726–4
1. Cable structures. 2. Shells (Engineering). I. Title. II. Series
TA660.C3V54 1990
624.1′774–dc20

90–43934
CIP

Table of contents

Introduction 7

1 The configuration of cable nets and tensegric shells 9
1.1 The configuration of cable nets 9
1.2 The configuration of tensegric shells 12

2 Conventional reticulated structures 20
2.1 The equilibrium matrix 20
2.2 The deformation matrix 22
2.3 The relationship between the equilibrium and the deformation matrices 25
2.4 The stiffness matrix 27
2.5 The analysis of conventional reticulated structures 28

3 Prestressing of cabled structures 33
3.1 The importance of prestressing 33
3.2 The feasibility of prestressing 33
3.3 The forces induced by prestressing 53
3.4 The deformation due to prestressing 57

4 The effect of external load 72
4.1 A fully constrained cabled structure 72
4.2 An under-constrained cabled structure: the fitted-load case 73
4.3 An under-constrained cabled structure: the non-fitted-load case 83

5 The dynamical behaviour of cabled structures 97
5.1 A fully constrained cabled structure: free vibration 97
5.2 A fully constrained cabled structure: forced vibration 103
5.3 An under-constrained cabled structure: free vibration 106
5.4 An under-constrained cabled structure: forced vibration 113

6 The design of a cabled structure 115
6.1 The preliminary design and configuration of under-constrained structures 115
6.2 The prestressable configuration 115
6.3 Constraints on the designed configuration 118

Appendix I The theory of a single cable 121

Suggestions for further reading 127

Index 129

Introduction

Cable nets are a reticulated assembly of cables suitable for the roofing of long spans without intermediate supports. In many cases, cable nets are the only feasible solution which often provides an economical roofing and presents a popular structure which has been extensively employed during the last 30 years.

Tensegric shells are a reticulated assembly of cables and bars. At each node, one bar only is connected to the cables.

There is a considerable similarity in the behaviour of and the problems concerning tensegric shells and cable nets. They may be identified as one type of structure, cabled structures.

The theory of cabled structures has been presented to the designer in a complicated form. A wide knowledge of the behaviour of geometrically non-linear structures is required. It is ranging far beyond the scope of the mundane knowledge of a structural engineer. Analysis is carried out by the use of intricate computer programs. These make comprehension and understanding of the behaviour of these structures very difficult and laborious.

The method introduced in this book shows how cabled structures can be analysed by following the general methods used to analyse conventional reticulated structures. The structural engineer is accustomed to these methods and uses them regularly.

1

The configuration of cable nets and tensegric shells

1.1 THE CONFIGURATION OF CABLE NETS

The simplest way to construct a cable net roof is to support the cables one parallel to the other between two edge elements as shown in Fig. 1.1.1. Because of the nature of

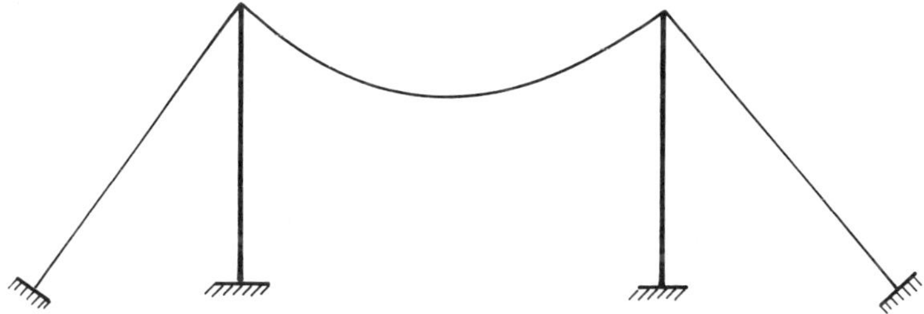

Fig. 1.1.1 — A simple cable roof.

the cables this roof does not have a fixed geometry. (The reader can refer to Appendix I where the theory of a single cable is presented and its properties are discussed.) In some cases this roof undergoes minor changes in its configuration due to a change in the applied load. For example, in the case where a point distributed load applies in addition to a uniformly distributed load, the cable net takes the form shown in Figs 1.1.2(a) and 1.1.2(b). A change in the location of the point distributed load causes minor changes in the cable configuration.

The minor changes in the cable configuration cause problems with the cladding but generally can be tolerated. An excessive wind uplift changes the configuration of

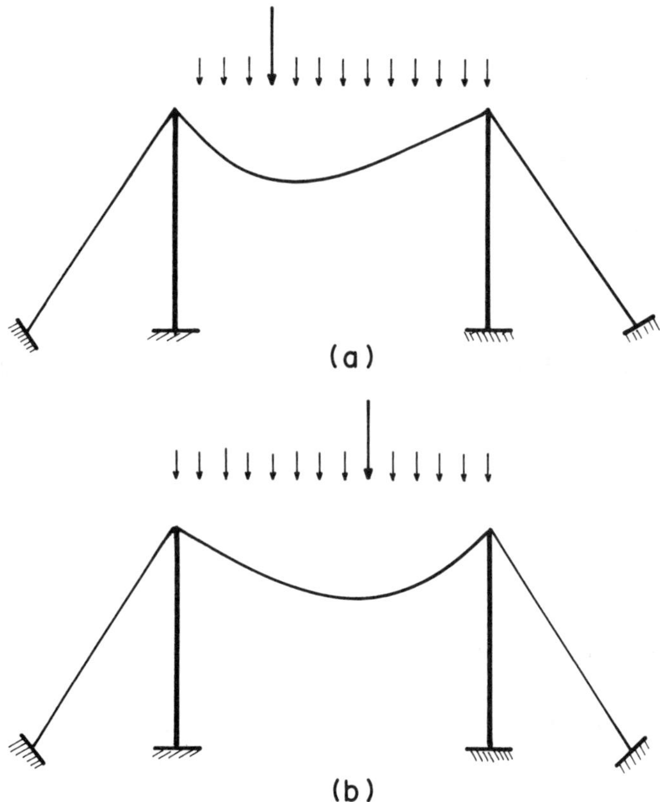

Fig. 1.1.2 — The change in the configuration of the cables.

the cables completely as shown in Fig. 1.1.3. This large change in the configuration can damage the cover of the structure and cause an uneasy feeling, even panic, in people sheltering in this structure.

One way to overcome the major change in the configuration of the cables is by increasing the self-weight of the roof so that there is an excessive downward load in the cases of the expected wind uplift. This method was used in early cable roofs. It seems wrong to increase the load for the sole purpose of opposing the wind uplift, and other methods were sought subsequently.

One method was to introduce another cable with curvature in the opposite direction as shown in Fig. 1.1.4. In this case, cables A sustain downward loads and cables B sustain the uplift. In order to put both types of cable into action in all loading cases, tension is induced into them by prestressing. The two types of cable are prestressed against each other. This is done by shorting cable A, cable B or the vertical cables. In this way, cables A and B sustain the downward loads as well as the uplift. A downward load increases the tension in cables A and reduces the tension in cables B. An uplift increases the tension in cables B and reduces it in cables A. In

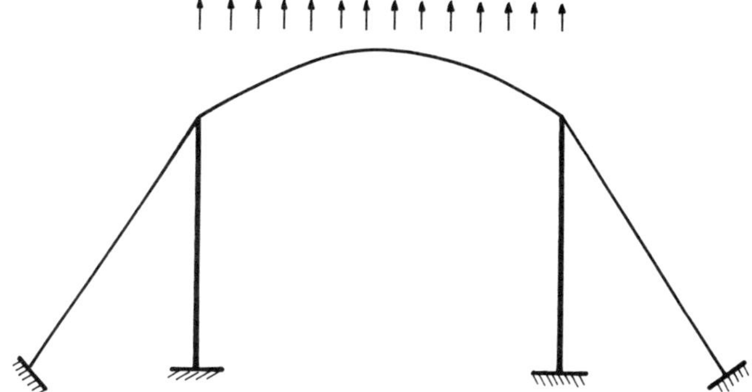

Fig. 1.1.3 — Large change in the configuration of the cables.

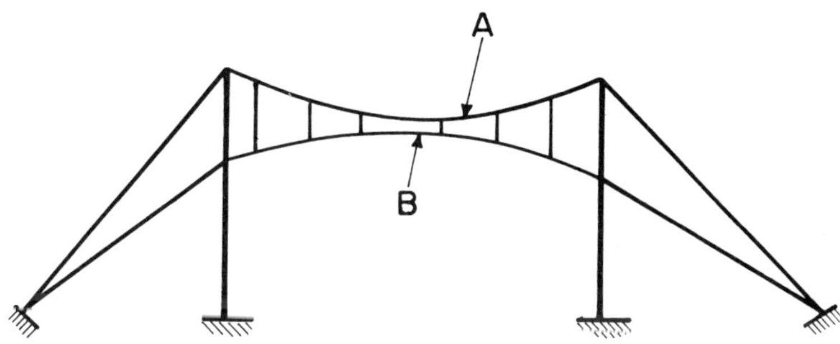

Fig. 1.1.4 — The two-cable system.

order to save the space lost between the two types of cable, they are arranged in space to form a single envelope, e.g. to form the saddle roofs shown in Fig. 1.1.5. The cables similar to cables A and cables B in Fig. 1.1.4 are indicated. In some cases the net is supported by a central mast as shown in Fig. 1.1.6. In this case the two families of cables are indicated by C and D. The net can be supported by a central arch as shown in Fig. 1.1.6. The two families of cables are indicated by A and B. In all these cases it is common practice to prestress the cable net by shorting one of the families of cables or by lengthening the edge elements.

The advantages of cable nets lie in their simplicity and the elimination of stability problems. Their major disadvantages are associated with their limited range of geometrical configurations. The possible geometries are awkward; in most cases the high points in the structure are near the supports where they are of minor benefit. For example, in the case of a dome the high points are at the centre where they are most advantageous, whereas in the case of the cable net shown in Fig. 1.1.5 they are at the edges. In order to obtain high points at the centre, the net is supported by an arch as shown in Fig. 1.1.7. The solution used in the case of the cable net shown in Fig. 1.1.6

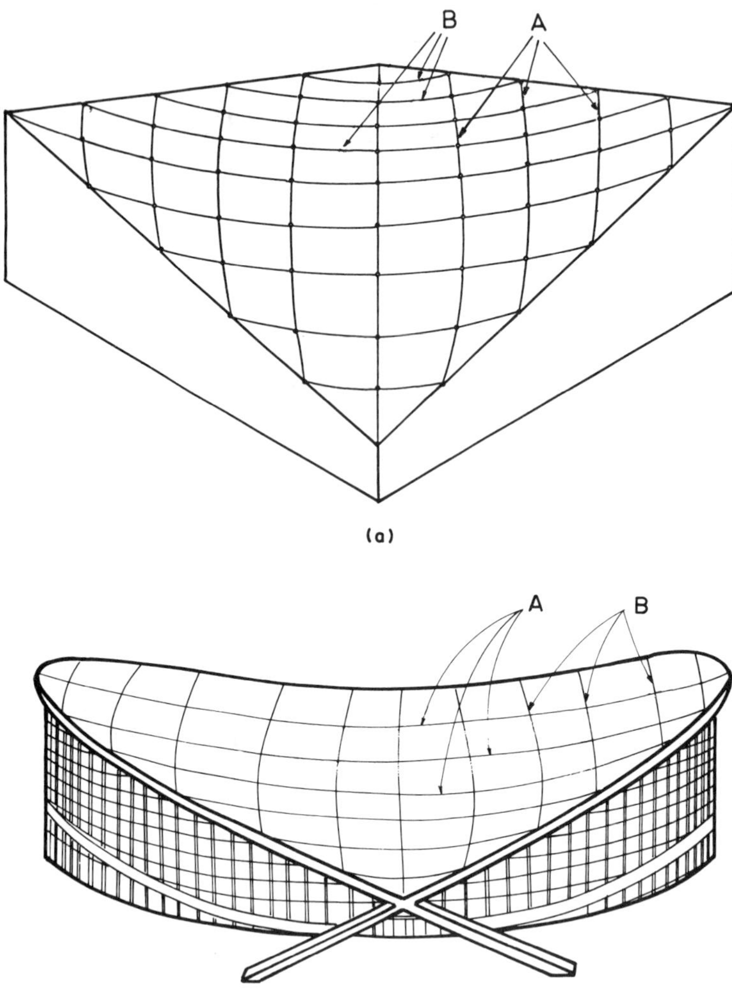

Fig. 1.1.5 — Saddle roofs.

is in most cases unsatisfactory, because the centre is occupied by the mast. As in most cases the high points of the roof are at the supports, the edge elements there are at a high level. This implies large edge elements in order to transfer the large forces induced into the edge elements to ground level.

1.2 THE CONFIGURATION OF TENSEGRIC SHELLS

A tensegric shell is a three-dimensional assembly of linear members. The shell consists of a cable net and compression bars with each of their ends connected to the nodes of the net. The net is stabilized and the shell gains its final shape by prestressing

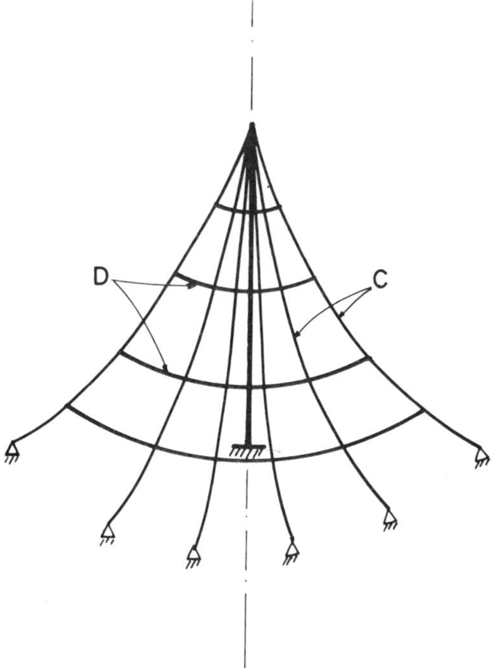

Fig. 1.1.6 — A cable net supported by a central mast.

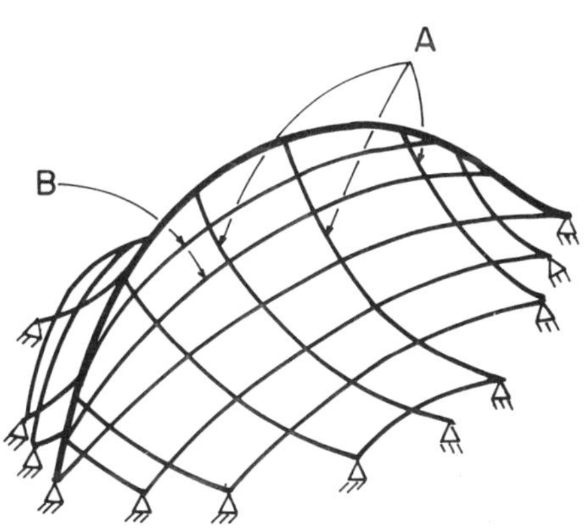

Fig. 1.1.7 — A net supported by an arch.

the bars against the cable net. Prestressing can take place by the lengthening of some of the bars or by the shortening of some of the cables. The bars are arranged in such a way that, when the shell gains its prestressed configuration, no bar touches any other, nor is connected to any other at a node. A spherical tensegric shell is shown in Fig. 1.2.1 and a tensegric shell in the shape of a cooling tower is shown in Fig 1.2.2.

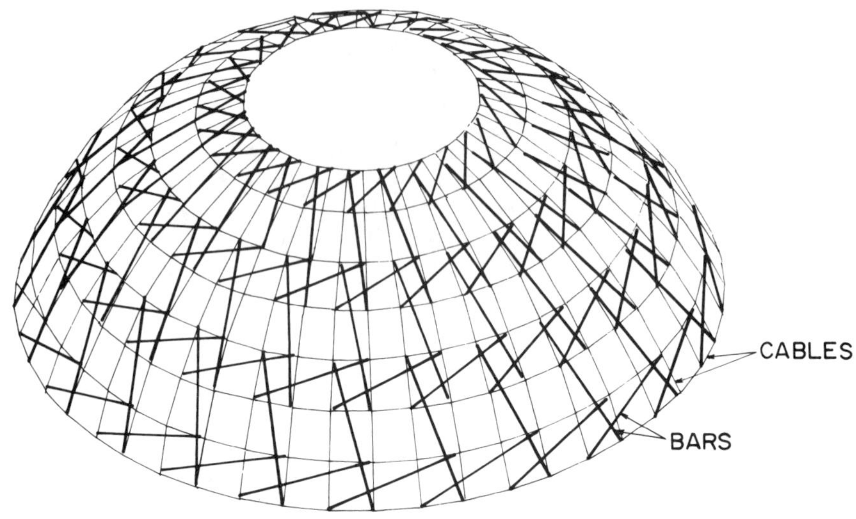

Fig. 1.2.1 — A spherical tensegric shell.

An artist's impression of a swimming pool covered by a spherical tensegric shell is shown in Fig. 1.2.3.

Tensegric shells are constructed from tensegric nets which are cable nets in which bars are fitted between the nodes of the net, one bar at a node. The bars are arranged in a typical way in which they cross each other but no bar is connected to another directly. A typical arrangement of the bars in the case of various cable nets is shown in Fig. 1.2.4. In Figs 1.2.4(a) and 1.2.4(b) the arrangements of the bars in the case of square and triangular cable nets, respectively, are shown. A similar method is used to construct the tensegric nets shown in Figs 1.2.4(c) and 1.2.4(d). These nets can be fitted to any required geometry of a shell. In Figs 1.2.1 and 1.2.2 the net shown in Fig. 1.2.4(a) is fitted into a dome and into a cooling tower shape, respectively. In Fig. 1.2.5 the net shown in Fig. 1.2.4(b) is fitted into a shallow dome and in Fig. 1.2.6 the net shown in Fig. 1.2.4(d) is fitted into a cylindrical shell. It can be seen that, where the tensegric net is fitted to a required configuration, the bars are connected at each of their ends to the nodes of the cable net and no bar touches any other.

Tensegric shells are a new type of structure and no large tensegric shell has been constructed yet. There are many technical advantages in tensegric shells which can be employed to overcome the problems associated with long-span structures.

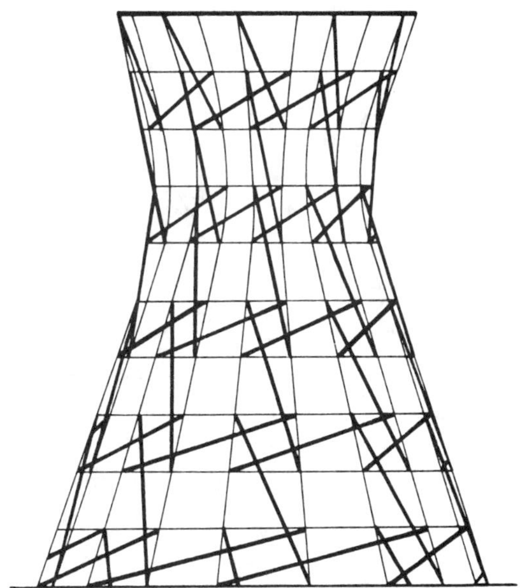

Fig. 1.2.2 — A tensegric shell in the shape of a cooling tower.

Tensegric shells are cost effective in both material and labour. Fabrication of the element reduces cost and enhances ease in transport. The same type of connector is used throughout the structure. (A possible simple connector is shown in Fig. 1.2.7.) The foundations of the shell are simple and no anchorage to the soil is required. Because the shell is stabilized by prestressing, its erection is simple in that no scaffolding is required, no special accuracy is needed and unskilled labour can be used. Tensegric shells are demountable which means that they can be used as temporary shelters.

Comparison of Figs 1.1.5, 1.1.6 and 1.1.7 with Figs 1.2.1, 1.2.5 and 1.2.6 shows that the space confined by tensegric shells is more suitable for practical usage than the space confined by cable nets. In cable nets, much space of awkward shape is wasted along the edges. In most cases the edge elements of cable nets are placed at high levels and they have to be of large dimensions to transfer the forces acting on them to ground level. In tensegric shells, no prominent edge elements are required.

The bars of tensegric shells are more complicated to manufacture than the simple elements of cable nets. The construction of tensegric shells is more difficult than the construction of cable nets, especially at the stage where the bars are placed in their proper locations.

There is a great similarity between the behaviour of cable nets and the behaviour of tensegric shells. Because of this similarity the two types of structure are considered to belong to a single family of structures identified as **cabled structures**. In the following pages the problem associated with the behaviour and the analysis of structures of this family are investigated and discussed.

Fig. 1.2.3 — An artist's impression of a swimming pool covered by a spherical tensegric shell

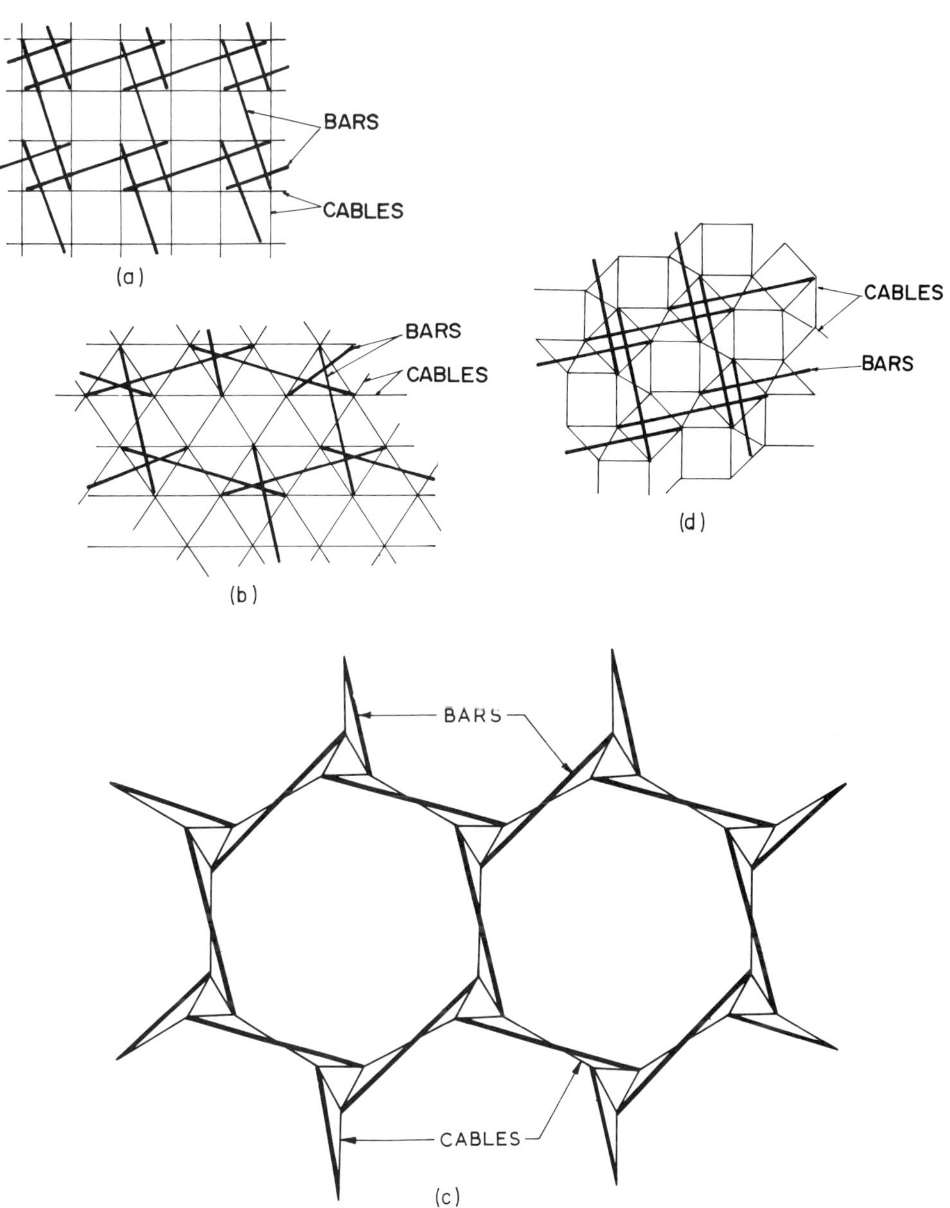

Fig. 1.2.4 — Various tensegric nets.

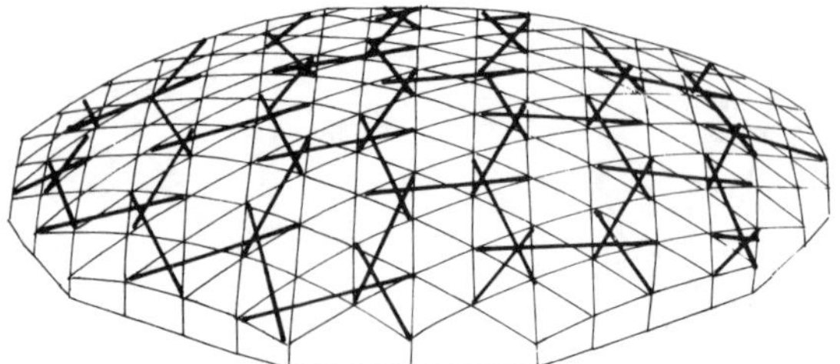

Fig. 1.2.5 — A shallow spherical tensegric shell.

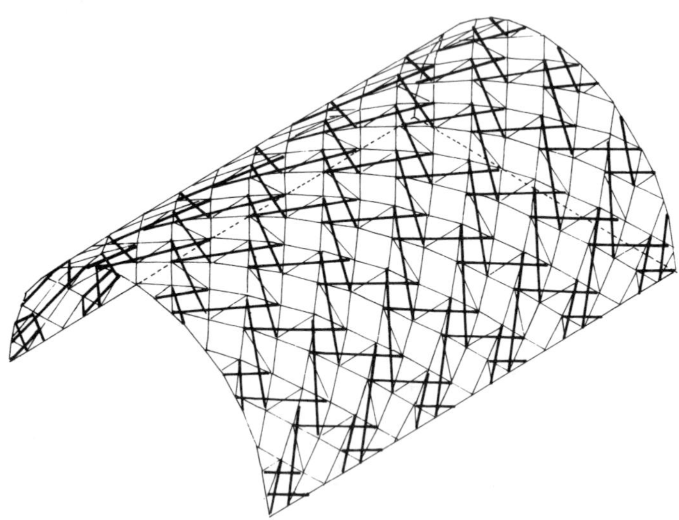

Fig. 1.2.6 — A tensegric cylindrical shell.

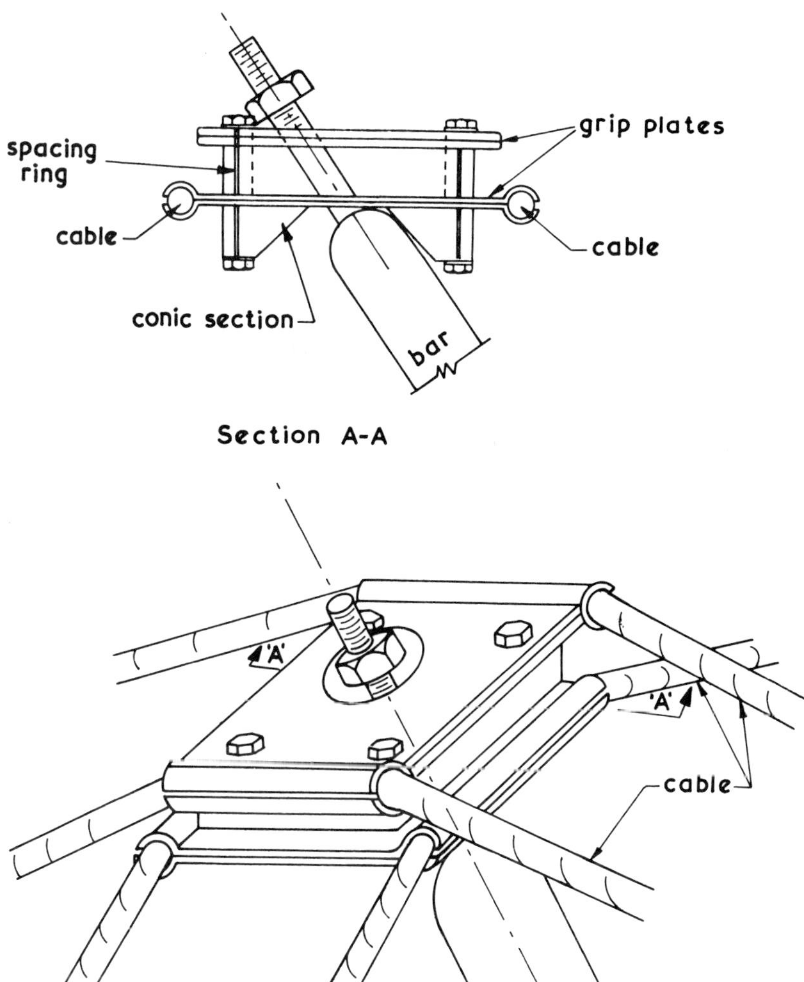

Fig. 1.2.7 — A possible connector in tensegric shells.

2

Conventional reticulated structures

2.1 THE EQUILIBRIUM MATRIX

The external load and internal forces acting at node j of a two-dimensional reticulated structure take the form shown in Fig. 2.1.1. The external load acting at

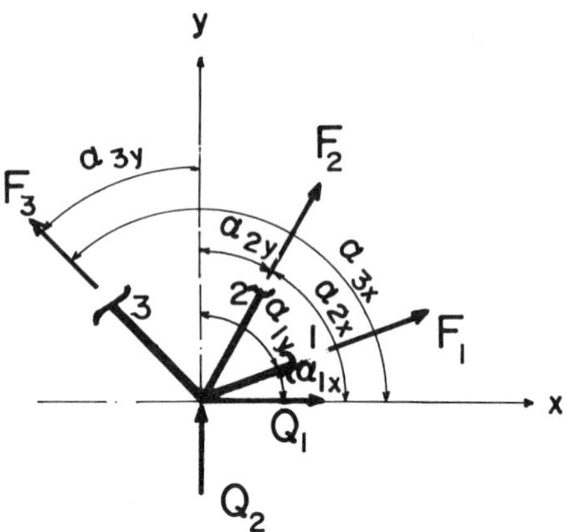

Fig. 2.1.1 — The forces at joint j.

the nodes is presented by its components acting in the x and y directions. For the node shown in Fig. 2.1.1, it is described by Q_1 and Q_2. Q_1 and Q_2 are assumed to be positive when acting in the positive direction of the relevant axes.

The internal force at member i is assumed positive in the case of internal tension and negative in the case of internal compression. In the process of analysis they are assumed to be positive first. The sign of the solution, positive or negative, indicates their true nature. α_{ix} is the angle between the direction of the positive internal force acting at member i and the positive direction of the x axis. α_{iy} is the angle between the direction of the positive internal force acting at member i and the positive direction of the y axis. In the two-dimensional case the geometry of the bar configuration implies that

$$\cos \alpha_{ix} = \sin \alpha_{iy} \tag{2.1.1}$$

which in the three-dimensional case takes the form

$$\cos^2 2\alpha_{ix} + \cos^2 \alpha_{iy} + \cos^2 \alpha_{iz} = 1 \ . \tag{2.1.2}$$

α_{iz} is the angle between the direction of the positive external force acting at member i and the positive direction of the z axis. The equilibrium condition of statics at node j in the x direction implies that

$$Q_1 = -F_1 \cos \alpha_{1x} - F_2 \cos \alpha_{2x} - F_3 \cos \alpha_{3x} \ . \tag{2.1.3}$$

In the y direction it takes the form

$$Q_2 = -F_1 \cos \alpha_{1y} - F_2 \cos \alpha_{2y} - F_3 \cos \alpha_{3y} \ . \tag{2.1.4}$$

In the three-dimensional case a similar third equation, considering the equilibrium condition of statics at node j in the z direction, should be added.

For example, for the two-dimensional truss shown in Fig. 2.1.2, the equilibrium

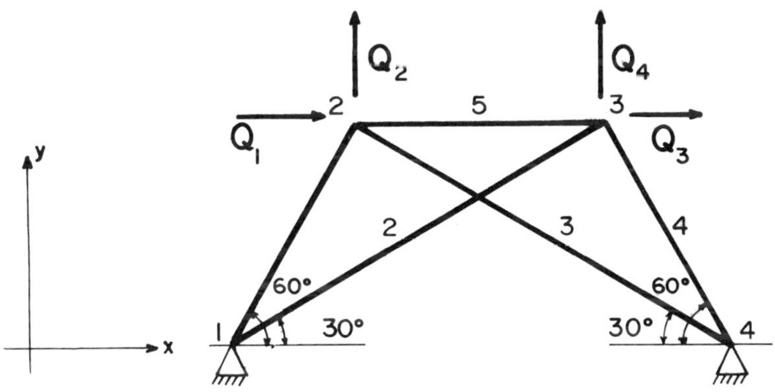

Fig. 2.1.2 — A truss.

conditions of statics, at the **free nodes** 2 and 3 (nodes not at the supports of the structure), take the form

$$Q_1 = \frac{1}{2} F_1 - \frac{\sqrt{3}}{2} F_3 - F_5$$

$$Q_2 = \frac{\sqrt{3}}{2} F_1 + \frac{1}{2} F_3 \tag{2.1.5}$$

$$Q_3 = \frac{\sqrt{3}}{2} F_2 - \frac{1}{2} F_4 + F_5$$

$$Q_4 = \frac{1}{2} F_2 + \frac{\sqrt{3}}{2} F_4 \, .$$

Equation (2.1.5) may be written in matrix form as

$$\mathbf{Q} = \mathbf{A}\mathbf{F}. \tag{2.1.6}$$

In this example, matrix **A** takes the form

$$\mathbf{A} = \begin{bmatrix} \frac{1}{2} & 0 & -\frac{\sqrt{3}}{2} & 0 & -1 \\ \frac{\sqrt{3}}{2} & 0 & \frac{1}{2} & 0 & 0 \\ 0 & \frac{\sqrt{3}}{2} & 0 & -\frac{1}{2} & 1 \\ 0 & \frac{1}{2} & 0 & \frac{\sqrt{3}}{2} & 0 \end{bmatrix} . \tag{2.1.7}$$

Vector **Q** is composed of the components of the external loads which act at all the free nodes. The number of elements of vector **Q** is equal to twice the number of free nodes in the two-dimensional case or to three times the number of free nodes in the three-dimensional case. The elements of vector **F** are the internal forces at the members of the structure. The size of vector **F** is the number of members of the structure. Matrix **A** which is based in the equilibrium conditions of statics only is called the **equilibrium matrix**.

In every reticulated structure the equilibrium conditions of statics at the free nodes take the form given by equation (2.1.6). In two-dimensional structures, equation (2.1.6) is found by considering the equilibrium conditions of statics in the given two directions at each free node. In three-dimensional cases, equation (2.1.6) includes also the third equilibrium condition of statics at each free node.

2.2 THE DEFORMATION MATRIX

The free nodes are free to move when the reticulated structure is subjected to external load. The degree of freedom in determining the nodal displacements is called the **kinematic degree of freedom**.

For the truss shown in Fig. 2.2.2, joints 2 and 3 are free nodes. The displacement of node 2 is determined by considering its displacement U_1 in the x direction and its displacement U_2 in the y direction as shown in Fig. 2.2.1. The displacement of node 3

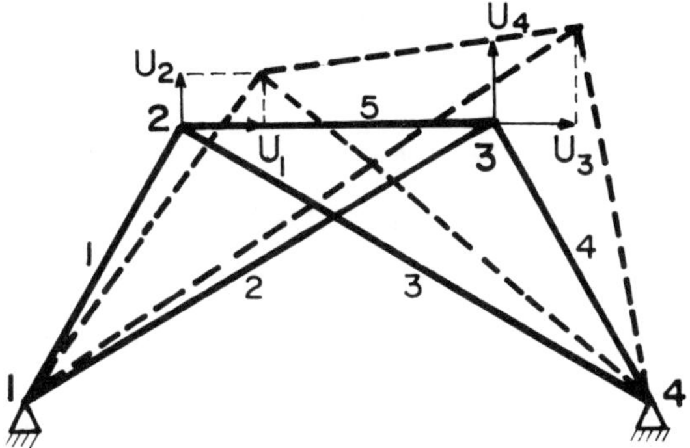

Fig. 2.2.1 — The deformed truss shown in Fig. 2.1.2.

is determined by U_3 and U_4. Since four displacements are required to determine the nodal displacements, the kinematic degree of freedom of the structure is 4.

By comparing Fig. 2.2.1 with Fig. 2.1.2 it can be seen that there is a one-to-one relationship between the components of the external loads and the possible nodal displacements. Q_1 is related to U_1, Q_2 to U_2, and so on. The number of components of the external loads is equal to the kinematic degrees of freedom of the structure.

The nodal displacements are associated with a change in the length of the members. The nodal displacements and the change in length of member i are shown in Fig. 2.2.2.

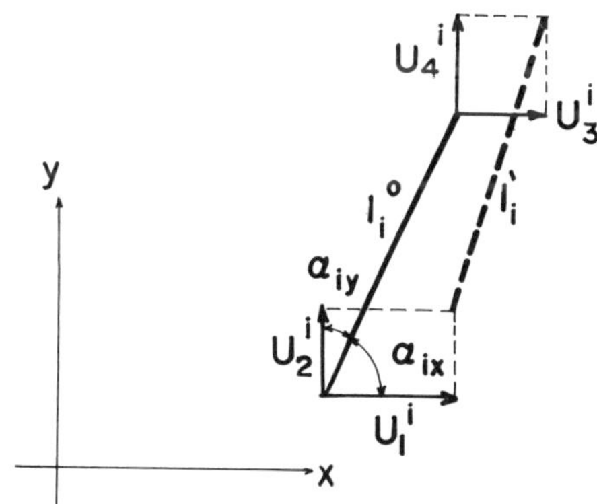

Fig. 2.2.2 — The displacement of bar i.

Because of the nodal displacements the length of member i changes from l_i^0 to l_i^1. The change Δ_i in length takes the form

$$\Delta_i = l_i^1 - l_i^0 \ . \tag{2.2.1}$$

By considering U_1^i, U_2^i, U_3^i and U_4^i, equation (2.2.1) takes the form

$$l_i^1 = \sqrt{[(l_i^0 \cos\alpha_{ix} + (U_3^i - U_1^i)]^2 + [l_i^0 \cos\alpha_{iy} + (U_4^i - U_2^i)]^2} \ . \tag{2.2.2}$$

Equation (2.2.2) may be written in the form

$$\Delta_i = l_i^0 \sqrt{1 + \frac{(2\cos\alpha_{ix})(U_3^i - U_1^i)}{l_i^0} + \frac{(2\cos\alpha_{iy})(U_4^i - U_2^i)}{l_i^0} + \frac{(U_3^i - U_1^i)^2}{(l_i^0)^2} + \frac{(U_4^i - U_2^i)^2}{(l_i^0)^2}} - l_i^0 \ . \tag{2.2.3}$$

For small nodal displacements, i.e.

$$\frac{U_3^i - U_2^i}{l_i^0} \approx \frac{U_4^i - U_2^i}{l_i^0} \ll 1 \ , \tag{2.2.4}$$

the fourth and fifth terms under the square root of equation (2.2.3) can be ignored in comparison with the second and third terms, and equation (2.2.3) takes the form

$$\Delta_i = l_i^0 \sqrt{1 + \frac{(2\cos\alpha_{ix})(U_3^i - U_1^i)}{l_i^0} + \frac{(2\cos\alpha_{iy})(U_4^i - U_2^i)}{l_i^0}} \ . \tag{2.2.5}$$

Equation (2.2.4) implies that the second and third terms under the square root of equation (2.2.5) are small in comparison with 1. By considering the fact that where ε is small,

$$\sqrt{1 + 2\varepsilon} = 1 + \varepsilon - \frac{\varepsilon^2}{2} \ldots \tag{2.2.6}$$

and using equation (2.2.6) where the second order is ignored, equation (2.2.5) takes the form

$$\Delta_i = (\cos\alpha_{ix})(U_3^i - U_1^i) + (\cos\alpha_{iy})(U_4^i - U_1^i) \ . \tag{2.2.7}$$

For the truss shown in Fig. 2.2.1 the nodal displacements take the form

$$\begin{aligned}
\Delta_1 &= \frac{1}{2} U_1 + \frac{\sqrt{3}}{2} U_2 \\
\Delta_2 &= \frac{\sqrt{3}}{2} U_3 + \frac{1}{2} U_4 \\
\Delta_3 &= -\frac{\sqrt{3}}{2} U_1 + \frac{1}{2} U_2 \\
\Delta_4 &= -\frac{1}{2} U_3 + \frac{\sqrt{3}}{2} U_4
\end{aligned} \tag{2.2.8}$$

$$\Delta_5 = -U_1 + U_3 \ .$$

Equation (2.2.8) may be written in the matrix form

$$\mathbf{\Delta} = \mathbf{BU} \ . \tag{2.2.9}$$

In this example, matrix **B** takes the form

$$\begin{bmatrix} \frac{1}{2} & \frac{\sqrt{3}}{2} & 0 & 0 \\ 0 & 0 & \frac{\sqrt{3}}{2} & \frac{1}{2} \\ -\frac{\sqrt{3}}{2} & \frac{1}{2} & 0 & 0 \\ 0 & 0 & -\frac{1}{2} & \frac{\sqrt{3}}{2} \\ -1 & 0 & 1 & 0 \end{bmatrix} . \tag{2.2.10}$$

Matrix **B** is called the **deformation matrix**.

2.3 THE RELATIONSHIP BETWEEN THE EQUILIBRIUM AND THE DEFORMATION MATRICES

The relationship between matrices **A** and **B** can be established by using the **principle of virtual work**. The principle of virtual work states that the work performed by a system of forces in equilibrium in going through any displacement is zero. The validity of this principle is easy to observe. When the system is in equilibrium, its resultant is equal to zero. The work done by a zero force in going through any displacement is obviously zero.

Consider the case in which the structure shown in Fig. 2.1.2 is loaded in such a way that a force F_2 is induced in member 2 and all the rest of the internal forces are equal to zero; then $F_1 = F_2 = F_4 = F_5 = 0$. The loads in this case can be found by using equation (2.1.6) which takes the form

$$\begin{aligned} Q_1 &= A_{12}F_2 \\ Q_2 &= A_{22}F_2 \\ Q_3 &= A_{32}F_2 \\ Q_4 &= A_{42}F_2 \end{aligned} \tag{2.3.1}$$

in which A_{ij} is the element ij of matrix **A**. Consider the virtual work of these loads and forces associated with the nodal displacement in which node 3 moves U_4 in the y direction only. By using equation (2.2.9) the change in length of the members takes the form

$$\begin{aligned}\Delta_1 &= B_{14}U_4\\ \Delta_2 &= B_{24}U_4\\ \Delta_3 &= B_{34}U_4\\ \Delta_4 &= B_{44}U_4\\ \Delta_5 &= B_{54}U_4\end{aligned} \tag{2.3.2}$$

in which B_{ij} is element ij of matrix **B**.

The load, internal forces and the nodal displacements at node 3 take the form shown in Fig. 2.3.1. The virtual work at node 3 implies that

$$Q_4U_4 + Q_3U_3 - F_4\Delta_4 - F_2\Delta_2 - F_5\Delta_5 = 0 \ . \tag{2.3.3}$$

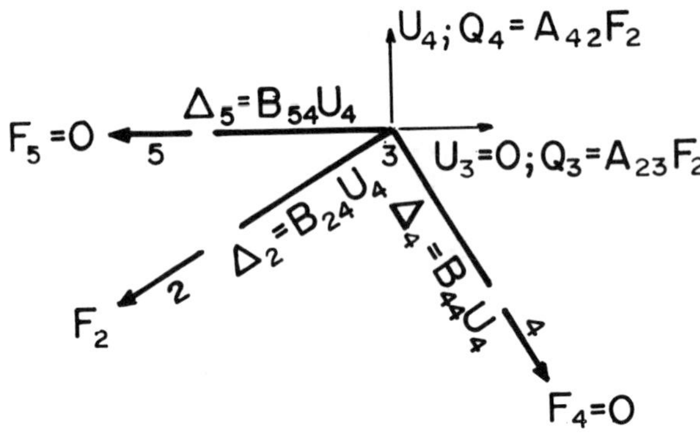

Fig. 2.3.1 — The displacement and forces at node 3.

By using equations (2.3.1) and (2.3.2), equation (2.3.3) takes the form

$$A_{42}F_2U_4 - F_2B_{24}U_4 = 0 \tag{2.3.4}$$

which implies that

$$A_{42} = B_{24} \ . \tag{2.3.5}$$

Equation (2.3.5) indicates that, in general,

$$A_{ij} = B_{ji} \tag{2.3.6}$$

which implies the following relationship between matrices **A** and **B**:

$$\mathbf{B} = \mathbf{A}^{\mathrm{T}} \ . \tag{2.3.7}$$

By using equation (2.3.7), equation (2.2.9) takes the form

$$\mathbf{\Delta} = \mathbf{A}^{\mathrm{T}}\mathbf{U} \ . \tag{2.3.8}$$

2.4 THE STIFFNESS MATRIX

The change Δ_i in the length of member i is related to the internal force in the member. In the case where the member is made of an elastic material, this relationship is given by Hooke's law

$$F_i = \frac{E_i A_i}{l_i^0} \Delta_i \tag{2.4.1}$$

in which E_i is the modulus of elasticity of the material of member i, A_i is the area of its cross-section and l_i^0 is its length. For the truss shown in Fig. 2.1.2, equation (2.4.1) takes the form

$$\begin{aligned}
F_1 &= \frac{E_1 A_1}{l_1^0} \Delta_1 \\
F_2 &= \frac{E_2 A_2}{l_2^0} \Delta_2 \\
F_3 &= \frac{E_3 A_3}{l_3^0} \Delta_3 \\
F_4 &= \frac{E_4 A_4}{l_4^0} \Delta_4 \\
F_5 &= \frac{E_5 A_5}{l_5^0} \Delta_5 \ .
\end{aligned} \tag{2.4.2}$$

Equation (2.4.2) may be written in the matrix form

$$\mathbf{F} = \mathbf{S}\mathbf{\Delta} \ . \tag{2.4.3}$$

S is the **stiffness matrix** which is a **square mtrix**, where the number of columns is equal to the number of rows, and a **diagonal matrix**, where all elements except those on the principal diagonal are equal to zero. For the truss shown in Fig. 2.1.2, matrix **S** takes the form

$$\mathbf{S}=\begin{bmatrix} \frac{E_1A_1}{l_1} & 0 & 0 & 0 & 0 \\ 0 & \frac{E_1A_1}{l_1} & 0 & 0 & 0 \\ 0 & 0 & \frac{E_1A_1}{l_1} & 0 & 0 \\ 0 & 0 & 0 & \frac{E_1A_1}{l_1} & 0 \\ 0 & 0 & 0 & 0 & \frac{E_1A_1}{l_1} \end{bmatrix} . \tag{2.4.4}$$

2.5 THE ANALYSIS OF CONVENTIONAL RETICULATED STRUCTURES

By using equations (2.1.6), (2.2.9), (2.3.7) and (2.4.3) the internal forces induced by the external load, the change in the length of the members and the nodal displacements of reticulated structures can be analysed:

$$\begin{aligned} \mathbf{Q} &= \mathbf{AF} \\ \mathbf{F} &= \mathbf{S\Delta} \\ \mathbf{\Delta} &= \mathbf{A}^{\mathrm{T}}\mathbf{U} \ . \end{aligned} \tag{2.5.1}$$

In the case of a **determinate structure** the number of unknown internal forces induced in the members is equal to the number of equilibrium equations. In this case, matrix **A** is a square matrix and the internal forces can be found directly from the first equation of equation (2.5.1) by using matrix **A** only:

$$\mathbf{F} = \mathbf{A}^{-1}\mathbf{Q} \ . \tag{2.5.2}$$

By using **F**, the nodal displacements can be found by using the second and third equations of equation (2.5.1):

$$\mathbf{U} = (\mathbf{SA}^{\mathrm{T}})^{-1}\mathbf{F} \ . \tag{2.5.3}$$

For example, the truss shown in Fig. 2.5.1 is a determinate structure. Matrix **A** takes the form

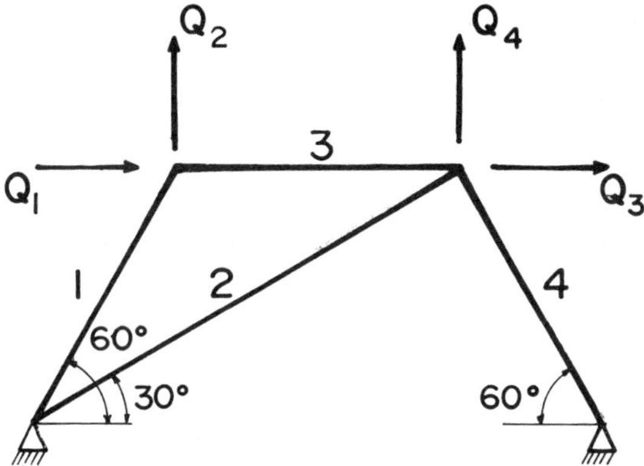

Fig. 2.5.1 — A determinate structure.

$$\mathbf{A} = \begin{bmatrix} \frac{\sqrt{3}}{2} & 0 & -1 & 0 \\ \frac{1}{2} & 0 & 0 & 0 \\ 0 & \frac{\sqrt{3}}{2} & 1 & -\frac{1}{2} \\ 0 & \frac{1}{2} & 0 & \frac{\sqrt{3}}{2} \end{bmatrix} . \quad (2.5.4)$$

In this case, **A** is a square matrix and the forces in the members can be analysed by using equation (2.5.2) without considering the stiffness matrix. The stiffness matrix is used to determine the nodal displacements by the employment of equation (2.5.3).

In the case of an **indeterminate structure** there are more unknown internal forces in the members than equilibrium equations. In this case, it is impossible to determine the internal forces by using the first equation of equation (2.5.1) only. Only by considering the siffness matrix is it possible to determine the change in the internal forces and the nodal displacement. In this case, equation (2.5.1) takes the form

$$\begin{aligned} \mathbf{F} &= \mathbf{S}\mathbf{A}^{\mathrm{T}}\mathbf{D}^{-1}\mathbf{Q} \\ \mathbf{U} &= \mathbf{D}^{-1}\mathbf{Q} \\ \mathbf{D} &= \mathbf{A}\mathbf{S}\mathbf{A}^{\mathrm{T}} . \end{aligned} \quad (2.5.5)$$

D is the **generalized stiffness matrix**. For example, the truss shown in Fig. 2.1.2 is indeterminate. **A** is a 4×5 matrix. There are four equilibrium equations and five unknown internal forces. In this case it is impossible to determine the internal forces by equilibrium considerations only. The stiffness of the bars should be considered. The stiffness matrix is a 5 × 5 matrix. The first equation of equation (2.5.5) is a set of five equations associated with the four elements of the vector of the external load. The second equation is a set of four equations associated with the four elements of the vector of the external load. It should be noted that the generalized stiffness matrix (4 × 4) is a square matrix with a feasible transverse.

In the case of a **mechanism**, there are more equilibrium equations than unknown internal forces. There are more rows than columns in matrix **A**. The modal displacements given by the third equation of equation (2.5.1) are not uniquely defined by the change in length of the members. It is possible to change the configuration of the structure without changing the length of the members.

For example, the structure shown in Fig. 2.5.2. is composed of three bars and two

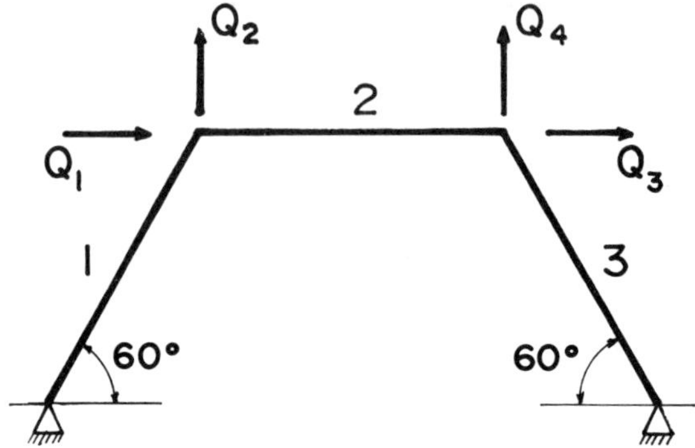

Fig. 2.5.2 — A mechanism.

free nodes. Matrix **A** takes the form

$$\mathbf{A} = \begin{bmatrix} \frac{1}{2} & -1 & 0 \\ \frac{\sqrt{3}}{2} & 0 & 0 \\ 0 & 1 & -\frac{1}{2} \\ 0 & 0 & \frac{\sqrt{3}}{2} \end{bmatrix}. \tag{2.5.6}$$

By using equation (2.5.6) the third equation of equation (2.5.1) takes the form

$$\Delta_1 = \frac{1}{2}U_1 + \frac{\sqrt{3}}{2}U_2$$

$$\Delta_2 = -U_1 + U_3 \qquad (2.5.7)$$

$$\Delta_3 = -\frac{1}{2}U_3 + \frac{\sqrt{3}}{2}U_4 .$$

It can be seen that for a certain change in length of the members it is necessary to assume the magnitude of U_4 and only then the magnitudes of U_1, U_2 and U_3 are uniquely defined. Even in the case where $\Delta_1 = \Delta_2 = \Delta_3 = 0$, it is possible to determine U_1, U_2 and U_3 for a given U_4. This is the mathematical formulation of the trivial fact that the mechanism shown in Fig. 2.5.2 has no **fixed geometry**. It can change its geometrical configuration without distorting its members as shown in Fig. 2.5.3.

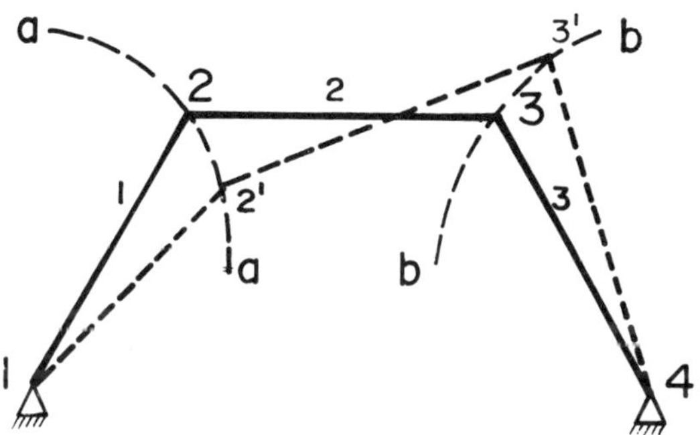

Fig. 2.5.3 — Possible distortion of the mechanism shown in Fig. 2.5.2.

Any points on curve aa and bb which are at a distance equal to the length of member 2 are possible locations of nodes 2 and 3. The members of the structure shown by the broken lines in Fig. 2.5.3 where the nodes at points 1, 2′, 3′ and 4 have the same lengths as the members of the structure with nodes at 1, 2, 3 and 4. The change in configuration is not associated with a change in the length of the members. The structure can change its configuration without the application of an external load. From a practical point of view such a mechanism is considered to be an unsatisfactory structure unsuitable for practical purposes. In the case where one is not aware of the fact that the nodal displacements are not uniquely defined, the method of analysis of indeterminate structures can be followed and equation (2.5.5) can be used. In this case the stiffness matrix is used to determine matrix **D**. Because of the nature of matrix **A**, matrix **D** is a **singular matrix** with a determinate equal to zero. In this case the second equation of equation (2.5.5) is not valid, which indicates again that the nodal displacements are not uniquely defined.

In the case where all bars of the mechanism shown in Fig. 2.5.2 have the same cross-sectional areas and lengths matrix **S** takes the form

$$\mathbf{S} = \frac{E_0 A_0}{l} \begin{bmatrix} 1 & 0 & 0 \\ 0 & 1 & 0 \\ 0 & 0 & 1 \end{bmatrix}. \tag{2.5.8}$$

Matrix **D** takes the form

$$\mathbf{D} = \begin{bmatrix} \frac{5}{4} & \frac{\sqrt{3}}{4} & -1 & 0 \\ \frac{\sqrt{3}}{4} & \frac{3}{4} & 0 & 0 \\ -1 & 0 & \frac{5}{4} & -\frac{\sqrt{3}}{4} \\ 0 & 0 & -\frac{\sqrt{3}}{4} & \frac{3}{4} \end{bmatrix}. \tag{2.5.9}$$

It can be seen that the fourth row in matrix **D** is not independent; it is a linear combination of the first three rows:

$$D_{4j} = -\sqrt{3}D_{ij} + D_{2j} - \sqrt{3}D_{3j} \, . \tag{2.5.10}$$

The fact that one row can be seen as a linear combination of the other rows implies that the determinate of matrix **D** is equal to zero, matrix **D** is *singular* and the nodal displacements cannot be defined uniquely by using equation (2.5.5).

This is another method of indicating that the structure is a mechanism unsuitable for practical purposes.

3

Prestressing of cabled structures

3.1 THE IMPORTANCE OF PRESTRESSING

Like all practical structures, those structures that are used to cover large spans are required to preserve their geometrical configuration to a large extent. Only insignificant deformation of the structure in which the deformation of the members is small and does not draw attention is acceptable. Large distortion of the structure in which the elements of the structure undergo significant deformation causes an uneasy feeling in people sheltering in the structure and, in extreme cases, even panic. Thus, structures that undergo large distortions are unacceptable for practical purposes.

Where cables are used in the structure, it implies that the distortion of the structure should be small and that the cables should be always straight. To keep them straight, there should be a certain amount of tension in them all the time. Tension is induced in the cables by prestressing the structure. A feasible cable net or tensegric shell should satisfy the following two requirements.

(i) It should be prestressable.
(ii) Prestressing should induce tension in all cables.

Cable nets and tensegric shells that do not satisfy these two requirements are unacceptable and should not be used for practical purposes.

3.2 THE FEASIBILITY OF PRESTRESSING

In prestressing a cabled structure, forces are induced in its members. This is achieved either by shortening some of the cables or, in the case of tensegric shells, by also lengthening the bars.

For example, the simple two-cable nets shown in Fig. 3.2.1 can be prestressed by shortening some of the cables

For the simple tensegric shells shown in Fig. 3.2.2, it can be achieved by shortening some of the cables or by lengthening some of the bars.

The feasibility of prestressing a cabled structure is studied by examining the equilibrium conditions of statics at the free nodes. In the case when the external load is acting at these nodes, these equilibrium conditions take the matrix form given by equation (2.1.6). When the structure is prestressed and no external loads are acting on it, equation (2.1.6) takes the form

$$\mathbf{AP} = \mathbf{0} \,. \tag{3.2.1}$$

P is the vector of the forces induced in the members by prestressing. For a *fully constrained* cabled structure there are fewer equations in equation (3.2.1) denoted by m than unknown internal forces induced by prestressing denoted by n.

$$m < n, \text{ fully constrained structure.} \tag{3.2.2}$$

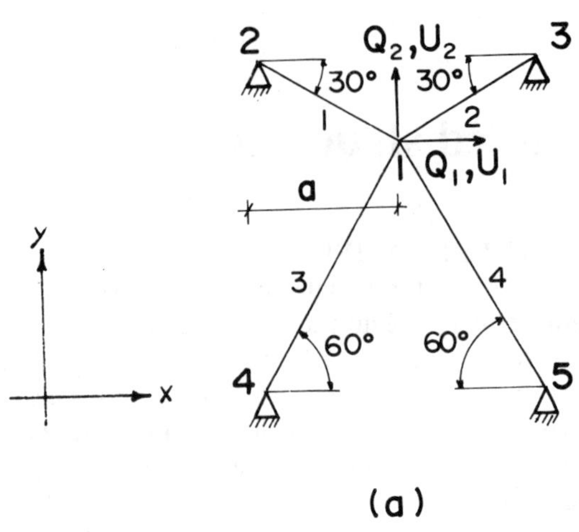

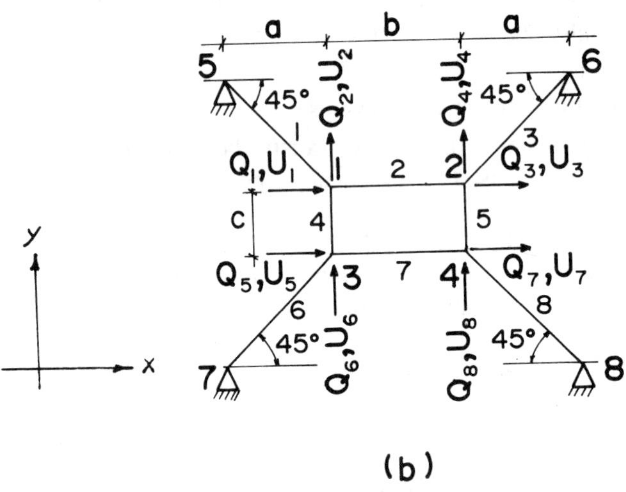

Fig. 3.2.1 — Cable nets.

For example the cable net shown in Fig. 3.2.1(a) and the tensegric shell shown in Fig. 3.2.2(a) are fully constrained cabled structures. In the case of the tensegric shell shown in Fig. 3.2.2(a), equation (3.2.1) takes the form

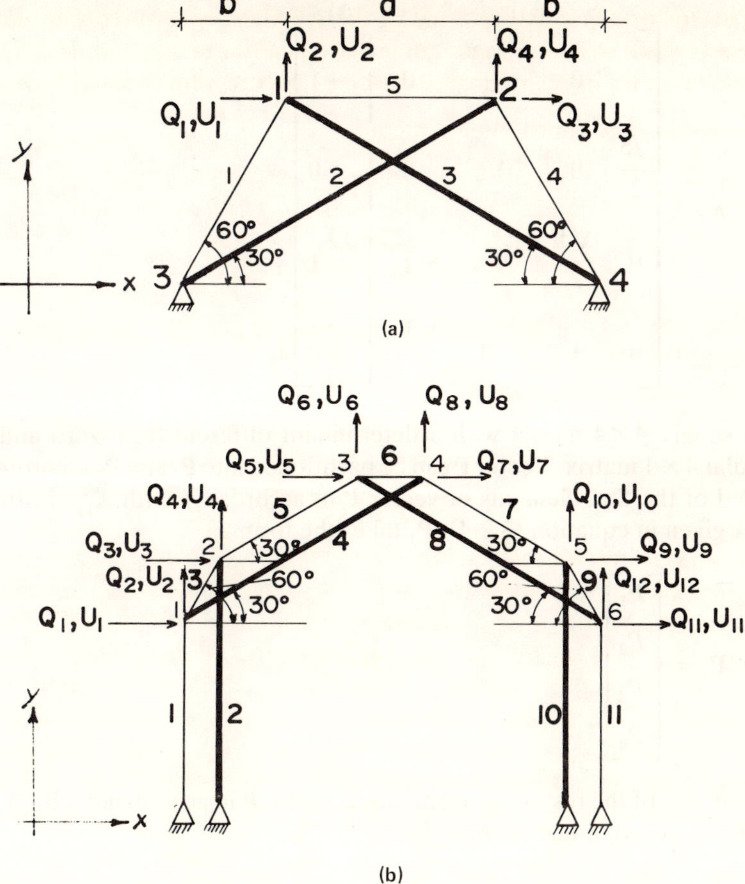

Fig. 3.2.2 — Tensegric shells.

$$
\begin{aligned}
&\tfrac{1}{2}P_1 - \frac{\sqrt{3}}{2}P_3 - P_5 = 0\\
&\frac{\sqrt{3}}{2}P_1 + \tfrac{1}{2}P_3 = 0\\
&\frac{\sqrt{3}}{2}P_2 - \tfrac{1}{2}P_4 + P_5 = 0\\
&\tfrac{1}{2}P_2 + \frac{\sqrt{3}}{2}P_4 = 0\,.
\end{aligned}
\tag{3.2.3}
$$

Matrix **A** in equation (3.2.4) can be partitioned into two parts: a square matrix $\mathbf{A}_{11}$ with a determinant different from zero and a rectangular matrix $\mathbf{A}_{12}$ composed appropriately of the rest of the elements of matrix **A**. For example matrix **A** of the equilibrium conditions of statics given by equation (3.2.3) can be partitioned into $\mathbf{A}_{11}$ and $\mathbf{A}_{12}$ as follows:

$$\mathbf{A} = \left[\begin{array}{cccc|c} \multicolumn{4}{c|}{\mathbf{A}_{11}} & \mathbf{A}_{12} \\ \frac{1}{2} & 0 & -\frac{\sqrt{3}}{2} & 0 & -1 \\ \frac{\sqrt{3}}{2} & 0 & \frac{1}{2} & 0 & 0 \\ 0 & \frac{\sqrt{3}}{2} & 0 & -\frac{1}{2} & 1 \\ 0 & \frac{1}{2} & 0 & \frac{\sqrt{3}}{2} & 0 \end{array}\right] . \quad (3.2.4)$$

$\mathbf{A}_{11}$ is a square 4×4 matrix with a determinant different from zero and $\mathbf{A}_{12}$ is a rectangular 4×1 matrix. Vector $\mathbf{P}$ can be partitioned into $\mathbf{P}_1$ and $\mathbf{P}_2$ accordingly. $\mathbf{P}_1$ is composed of the first elements of vector $\mathbf{P}$ in accordance with $\mathbf{A}_{11}$. In the case of matrix $\mathbf{A}$ given in equation (3.2.4), $\mathbf{P}_1$ takes the form

$$\mathbf{P}_1 = \begin{bmatrix} P_1 \\ P_2 \\ P_3 \\ P_4 \end{bmatrix} . \quad (3.2.5)$$

$\mathbf{P}_2$ is composed of the rest of the elements of vector $\mathbf{P}$ in accordance with $\mathbf{A}_{12}$. In this case it is composed of one element P_5:

$$\mathbf{P}_2 = [P_5] . \quad (3.2.6)$$

By using equations (3.2.4)–(3.2.6), equation (3.2.3) takes the form

$$\mathbf{A}_{11}\mathbf{P}_1 = -\mathbf{A}_{12}\mathbf{P}_2 . \quad (3.2.7)$$

By using equation (3.2.7) the forces induced in the members of the structure by prestressing can be determined. It is possible to assume P_2 and to determine $\mathbf{P}_1$ by using equation (3.2.7) which takes the form

$$\mathbf{P}_1 = -(\mathbf{A}_{11})^{-1}\mathbf{A}_{12}\mathbf{P}_2 . \quad (3.2.8)$$

The fact that it is possible to determine the prestressing forces implies that prestressing of this structure is feasible and requirement (i) is satisfied.

For the fully constrained cable net shown in Fig. 3.2.1(a), equation (3.2.1) takes the form

$$\begin{aligned} \frac{\sqrt{3}}{2} P_1 - \frac{\sqrt{3}}{2} P_2 + \tfrac{1}{2} P_3 - \tfrac{1}{4} P_4 &= 0 \\ -\tfrac{1}{2} P_1 - \tfrac{1}{2} P_2 + \frac{\sqrt{3}}{2} P_3 + \frac{\sqrt{3}}{2} P_4 &= 0 . \end{aligned} \quad (3.2.9)$$

In this case, $\mathbf{A}_{11}$, $\mathbf{A}_{12}$, $\mathbf{P}_1$ and $\mathbf{P}_2$ take the forms

$$\mathbf{A}_{11}=\begin{bmatrix}\frac{\sqrt{3}}{2} & -\frac{\sqrt{3}}{2}\\ -\frac{1}{2} & -\frac{1}{2}\end{bmatrix}, \qquad \mathbf{A}_{12}=\begin{bmatrix}\frac{1}{2} & -\frac{1}{2}\\ \frac{\sqrt{3}}{2} & \frac{\sqrt{3}}{2}\end{bmatrix} \tag{3.2.10}$$

$$\mathbf{P}_1=\begin{bmatrix}P_1\\ P_2\end{bmatrix}, \qquad \mathbf{P}_2=\begin{bmatrix}P_3\\ P_4\end{bmatrix}. \tag{3.2.11}$$

By using equation (3.2.8) it is possible to determine P_1 and P_2 by assuming P_3 and P_4. The fact that the forces induced in the members by prestressing can be determined implies that this cabled structure is prestressable and requirement (i) is satisfied.

In the uncommon cases where the determinants of all possible $\mathbf{A}_{11}$ are equal to zero, all possible $\mathbf{A}_{11}$ are singular and equation (3.2.8) cannot be used to determine the prestressing forces. In this case the cabled structure behaves as the under-constrained structure which is discussed below and the analysis of this type of cabled structure should be followed.

The cabled structures in which the number m of equations is equal to or larger than the number n of unknowns in equation (3.2.1) are *under-constrained* cabled structures:

$$m \geq n \;, \quad \text{under-constrained structure} \;. \tag{3.2.12}$$

For example, in the cable net shown in Fig. 3.2.1(b), $m=n=8$, the net is under-constrained and equation (3.2.1) takes the form

$$\begin{aligned}
&\frac{\sqrt{2}}{2}P_1-P_2=0\\
&-\frac{\sqrt{2}}{2}P_1+P_4=0\\
&P_2-\frac{\sqrt{2}}{2}P_3=0\\
&-\frac{\sqrt{2}}{2}P_3+P_5=0\\
&\frac{\sqrt{2}}{2}P_6-P_7=0\\
&\frac{\sqrt{2}}{2}P_6-P_4=0\\
&P_7-\frac{\sqrt{2}}{2}P_8=0\\
&-P_5+\frac{\sqrt{2}}{2}P_8=0\;.
\end{aligned} \tag{3.2.13}$$

In most cases there is only one trivial solution $P_1 = P_2 = P_3 = P_4 = P_5 = P_6 = P_7 = P_8 = 0$ to such a set of equations. In these cases it is not feasible to prestress the cable net and requirement (i) is not satisfied. However, there are cases where it is possible to find solutions different from the trivial one. To investigate this possibility, **Gaussian elimination** of matrix **A** is followed.

Matrix **A** of equation (3.2.13) takes the form

$$\mathbf{A} = \begin{bmatrix} \frac{\sqrt{2}}{2} & -1 & 0 & 0 & 0 & 0 & 0 & 0 \\ -\frac{\sqrt{2}}{2} & 0 & 0 & 1 & 0 & 0 & 0 & 0 \\ 0 & 1 & -\frac{\sqrt{2}}{2} & 0 & 0 & 0 & 0 & 0 \\ 0 & 0 & -\frac{\sqrt{2}}{2} & 0 & 1 & 0 & 0 & 0 \\ 0 & 0 & 0 & 0 & 0 & \frac{\sqrt{2}}{2} & -1 & 0 \\ 0 & 0 & 0 & -1 & 0 & -\frac{\sqrt{2}}{2} & 0 & 0 \\ 0 & 0 & 0 & 0 & 0 & 0 & 1 & -\frac{\sqrt{2}}{2} \\ 0 & 0 & 0 & 0 & -1 & 0 & 0 & \frac{\sqrt{2}}{2} \end{bmatrix}. \quad (3.2.14)$$

Before the elimination is started, the matrix is rearranged so that there will be large elements along the principal diagonal of the matrix. The rearrangement of matrix **A** takes the form

$$\mathbf{A} = \begin{bmatrix} -\frac{\sqrt{2}}{2} & 0 & 0 & 1 & 0 & 0 & 0 & 0 \\ \frac{\sqrt{2}}{2} & -1 & 0 & 0 & 0 & 0 & 0 & 0 \\ 0 & 1 & -\frac{\sqrt{2}}{2} & 0 & 0 & 0 & 0 & 0 \\ 0 & 0 & 0 & -1 & 0 & \frac{\sqrt{2}}{2} & 0 & 0 \\ 0 & 0 & -\frac{\sqrt{2}}{2} & 0 & 1 & 0 & 0 & 0 \\ 0 & 0 & 0 & 0 & 0 & \frac{\sqrt{2}}{2} & -1 & 0 \\ 0 & 0 & 0 & 0 & 0 & 0 & 1 & -\frac{\sqrt{2}}{2} \\ 0 & 0 & 0 & 0 & -1 & 0 & 0 & \frac{\sqrt{2}}{2} \end{bmatrix} . \tag{3.2.15}$$

The elimination is used to obtain all the terms below the principal diagonal matrix **A** equal to zero. At the first stage all the terms of the first column below the principal diagonal are reduced to zero by subtracting the appropriate multiples of the first row from each of the rows. In the case of matrix **A** given by equation (3.2.15), only the first element equal to $\sqrt{2}/2$ of the second row should be considered because the other elements of the first column below the principal diagonal are already equal to zero. In this case, after the first stage of the elimination, matrix **A** takes the form

$$\mathbf{A} = \begin{bmatrix} -\frac{\sqrt{2}}{2} & 0 & 0 & 1 & 0 & 0 & 0 & 0 \\ \boxed{0} & -1 & 0 & 1 & 0 & 0 & 0 & 0 \\ 0 & 1 & -\frac{\sqrt{2}}{2} & 0 & 0 & 0 & 0 & 0 \\ 0 & 0 & 0 & -1 & 0 & \frac{\sqrt{2}}{2} & 0 & 0 \\ 0 & 0 & -\frac{\sqrt{2}}{2} & 0 & 1 & 0 & 0 & 0 \\ 0 & 0 & 0 & 0 & 0 & \frac{\sqrt{2}}{2} & -1 & 0 \\ 0 & 0 & 0 & 0 & 0 & 0 & 1 & -\frac{\sqrt{2}}{2} \\ 0 & 0 & 0 & 0 & -1 & 0 & 0 & \frac{\sqrt{2}}{2} \end{bmatrix} . \tag{3.2.16}$$

At the second stage, the second row is used to bring all the elements of the second column below the principal diagonal to zero and so on until all elements below the principal diagonal are equal to zero. At the end of the elimination, matrix **A** takes the form

$$\mathbf{A} = \begin{bmatrix} -\frac{\sqrt{2}}{2} & 0 & 0 & 1 & 0 & 0 & 0 & 0 \\ 0 & -1 & 0 & 1 & 0 & 0 & 0 & 0 \\ 0 & 0 & -\frac{\sqrt{2}}{2} & 1 & 0 & 0 & 0 & 0 \\ 0 & 0 & 0 & -1 & 0 & \frac{\sqrt{2}}{2} & 0 & 0 \\ 0 & 0 & 0 & 0 & 1 & -\frac{\sqrt{2}}{2} & 0 & 0 \\ 0 & 0 & 0 & 0 & 0 & \frac{\sqrt{2}}{2} & -1 & 0 \\ 0 & 0 & 0 & 0 & 0 & 0 & 0 & -\frac{\sqrt{2}}{2} \\ 0 & 0 & 0 & 0 & 0 & 0 & 0 & 0 \end{bmatrix} . \tag{3.2.17}$$

It can be seen that in the last row all terms are equal to zero.

In a common case the eighth element of the eighth row is different from zero. When it is equal to zero, as in this example, it indicates that the determinant of matrix **A** is zero and the elements of the eighth row of matrix **A** given by equation (3.2.14) are dependent on the elements of the remaining seven rows and are a **linear combination** of them. The **rank** of matrix **A**, denoted by r, is equal to the number of rows in the matrix minus the number of dependent rows. In this case, $r = 8 - 1 = 7$, in which 1 indicates the fact that elements of one row are dependent. This implies that in equation (3.2.13) there are not eight independent equations but seven independent equations only. When these seven independent equations are satisfied, the eighth is satisfied automatically. In this case, matrix **A** can be partitioned into an indpendent part composed of the elements of the first seven rows and a dependent part composed of the elements of the last row which are associated with equilibrium in the U_8 direction. The independent part can be partitioned into a square 7×7 matrix $\mathbf{A}_{11}$ with a determinant different from zero and a rectangular 7×1 matrix $\mathbf{A}_{12}$ composed of the rest of the independent elements accordingly. The dependent part is partitioned into $\mathbf{A}_{21}$ (1×7) and $\mathbf{A}_{22}$ (1×1) accordingly. $\mathbf{A}_{11}$, $\mathbf{A}_{12}$, $\mathbf{A}_{21}$ and $\mathbf{A}_{22}$ take the forms

$$\mathbf{A} = \left[\begin{array}{ccccccc:c}
\multicolumn{7}{c}{\mathbf{A}_{11}} & \mathbf{A}_{12} \\
\frac{\sqrt{2}}{2} & -1 & 0 & 0 & 0 & 0 & 0 & 0 \\
-\frac{\sqrt{2}}{2} & 0 & 0 & 1 & 0 & 0 & 0 & 0 \\
0 & 1 & -\frac{\sqrt{2}}{2} & 0 & 0 & 0 & 0 & 0 \\
0 & 0 & -\frac{\sqrt{2}}{2} & 0 & 1 & 0 & 0 & 0 \\
0 & 0 & 0 & 0 & 0 & \frac{\sqrt{2}}{2} & -1 & 0 \\
0 & 0 & 0 & -1 & 0 & \frac{\sqrt{2}}{2} & 0 & 0 \\
0 & 0 & 0 & 0 & 0 & 0 & 1 & -\frac{\sqrt{2}}{2} \\
\hdashline
0 & 0 & 0 & 0 & -1 & 0 & 0 & \frac{\sqrt{2}}{2} \\
\multicolumn{7}{c}{\mathbf{A}_{21}} & \mathbf{A}_{22}
\end{array}\right] . \qquad (3.2.18)$$

$\mathbf{P}$ is partitioned accordingly into $\mathbf{P}_1$ (7×1) and $\mathbf{P}_2$ (1×1):

$$\mathbf{R} = \begin{bmatrix} P_1 \\ P_2 \\ P_3 \\ P_4 \\ P_5 \\ P_6 \\ P_7 \end{bmatrix} \tag{3.2.19}$$

$$\mathbf{P}_2 = [P_8] \ .$$

The Gaussian elimination given in equation (3.2.17) indicates that the elements of matrix $\mathbf{A}_{21}$ and of matrix $\mathbf{A}_{22}$ can be defined by using matrices $\mathbf{A}_{11}$ and $\mathbf{A}_{12}$ and a weighting vector $\mathbf{W}$:

$$\begin{aligned} \mathbf{A}_{21} &= \mathbf{W}\mathbf{A}_{11} \\ \mathbf{A}_{22} &= \mathbf{W}\mathbf{A}_{12} \ . \end{aligned} \tag{3.2.20}$$

By using equation (3.2.20), $\mathbf{W}$ takes the form

$$\mathbf{W} = (\mathbf{A}_{11})^{-1} \, \mathbf{A}_{21} \ . \tag{3.2.21}$$

In the case of matrix $\mathbf{A}$ given by equation (3.2.18), $\mathbf{W}$ takes the form

$$\mathbf{W} = [1 \quad 1 \quad 1 \quad -1 \quad -1 \quad 1 \quad -1] \ . \tag{3.2.22}$$

The forces induced in the members of the structure can be found by considering the independent part of matrix $\mathbf{A}$ only and by using equation (3.2.8). In this case, by using equation (3.2.8), P_1, P_2, P_3, P_4, P_5, P_6 and P_7 can be found as a function of P_8 by considering $\mathbf{A}_{11}$ and $\mathbf{A}_{12}$ given by equation (3.2.18). Because it is possible to determine the forces induced by the prestressing of the cable net shown in Fig. 3.2.1(b), prestressing of this cable net is feasible and requirement (i) is satisfied.

For the tensegric shell shown in Fig. 3.2.2(b), matrix **A** takes the form

$$
\mathbf{A} = \begin{bmatrix}
0 & 0 & -\frac{1}{2} & -\frac{\sqrt{3}}{2} & 0 & 0 & 0 & 0 & 0 & 0 & 0 \\
1 & 0 & -\frac{\sqrt{3}}{2} & -\frac{1}{2} & 0 & 0 & 0 & 0 & 0 & 0 & 0 \\
0 & 0 & \frac{1}{2} & 0 & -\frac{\sqrt{3}}{2} & 0 & 0 & 0 & 0 & 0 & 0 \\
0 & 1 & \frac{\sqrt{3}}{2} & 0 & -\frac{1}{2} & 0 & 0 & 0 & 0 & 0 & 0 \\
0 & 0 & 0 & 0 & \frac{\sqrt{3}}{2} & -1 & 0 & -\frac{\sqrt{3}}{2} & 0 & 0 & 0 \\
0 & 0 & 0 & 0 & \frac{1}{2} & 0 & 0 & \frac{1}{2} & 0 & 0 & 0 \\
0 & 0 & 0 & \frac{\sqrt{3}}{2} & 0 & 1 & -\frac{\sqrt{3}}{2} & 0 & 0 & 0 & 0 \\
0 & 0 & 0 & \frac{1}{2} & 0 & 0 & \frac{1}{2} & 0 & 0 & 0 & 0 \\
0 & 0 & 0 & 0 & 0 & 0 & \frac{\sqrt{3}}{2} & 0 & -\frac{1}{2} & 0 & 0 \\
0 & 0 & 0 & 0 & 0 & 0 & -\frac{1}{2} & 0 & \frac{\sqrt{3}}{2} & 0 & 0 \\
0 & 0 & 0 & 0 & 0 & 0 & 0 & \frac{\sqrt{3}}{2} & \frac{1}{2} & 0 & 0 \\
0 & 0 & 0 & 0 & 0 & 0 & 0 & -\frac{1}{2} & -\frac{\sqrt{3}}{2} & 1 & 0
\end{bmatrix} . \tag{3.2.23}
$$

In this case, equation (3.2.1) is composed of 12 equations, and 11 forces only are induced by prestressing. It seems that only where all the forces induced by the prestressing are equal to zero is equation (3.2.1) satisfied. If this is the case, it is impossible to prestress this tensegric shell and requirement (i) cannot be satisfied. In order to investigate the problem, Gaussian elimination is carried out.

The matrix is arranged so that the principal diagonal has terms different from zero. This is achieved by arranging the rows of matrix **A** in the following order 2, 4, 3, 1, 6, 5, 7, 11, 9, 12, 10 and 8 is the last row. At the end of the elimination, **A** takes the form

$$
\mathbf{A}=\begin{bmatrix}
1 & 0 & -\frac{\sqrt{3}}{2} & -1 & 0 & 0 & 0 & 0 & 0 & 0 & 0\\
0 & 1 & \frac{\sqrt{3}}{2} & 0 & -\frac{1}{2} & 0 & 0 & 0 & 0 & 0 & 0\\
0 & 0 & \frac{1}{2} & 0 & -\frac{\sqrt{3}}{2} & 0 & 0 & 0 & 0 & 0 & 0\\
0 & 0 & 0 & -\frac{\sqrt{3}}{2} & -\frac{\sqrt{3}}{2} & 0 & 0 & 0 & 0 & 0 & 0\\
0 & 0 & 0 & 0 & \frac{1}{2} & 0 & 0 & \frac{1}{2} & 0 & 0 & 0\\
0 & 0 & 0 & 0 & 0 & -1 & 0 & -\sqrt{3} & 0 & 0 & 0\\
0 & 0 & 0 & 0 & 0 & 0 & -\frac{\sqrt{3}}{2} & -\frac{\sqrt{3}}{2} & 0 & 0 & 0\\
0 & 0 & 0 & 0 & 0 & 0 & 0 & \frac{\sqrt{3}}{2} & \frac{1}{2} & 0 & 0\\
0 & 0 & 0 & 0 & 0 & 0 & 0 & 0 & 0 & 0 & 0\\
0 & 0 & 0 & 0 & 0 & 0 & 0 & 0 & -\frac{\sqrt{3}}{3}+\frac{1}{2\sqrt{3}} & 1 & 0\\
0 & 0 & 0 & 0 & 0 & 0 & 0 & 0 & 0 & 1 & 1\\
0 & 0 & 0 & 0 & 0 & 0 & 0 & 0 & 0 & 0 & 0
\end{bmatrix}.
\tag{3.2.24}
$$

It can be seen that the elements of two rows, 9 and 12, are all equal to zero. This indicates that the elements of these rows are a linear combination of the rest of the rows. The rank of matrix $\mathbf{A}$, $r = 12 - 2$, is 10. The ninth and twelfth rows of matrix $\mathbf{A}$ given by equation (3.2.24) correspond to the eighth and ninth rows of matrix $\mathbf{A}$ given by equation (3.2.23). The dependent part of matrix $\mathbf{A}$ given by equation (3.2.23) is composed of the elements of the eighth and ninth rows which are associated with the equilibrium condition in the U_8 and U_9 directions. The rest of the rows comprises the independent part. The independent part can be partitioned into a square 10×10 matrix $\mathbf{A}_{11}$ with a determinant different from zero and a rectangular 10×1 matrix $\mathbf{A}_{12}$ consisting of the rest of the independent elements accordingly. The dependent part is partitioned into $\mathbf{A}_{21}(2 \times 10)$ and $\mathbf{A}_{22}(2 \times 1)$ accordingly. $\mathbf{A}_{11}$, $\mathbf{A}_{12}$, $\mathbf{A}_{21}$ and $\mathbf{A}_{22}$ take the forms

$$\mathbf{A}_{11} \qquad\qquad \mathbf{A}_{12}$$

$$\mathbf{A} = \left[\begin{array}{cccccccccc|c}
0 & 0 & -\frac{1}{2} & -\frac{\sqrt{3}}{2} & 0 & 0 & 0 & 0 & 0 & 0 & 0 \\
1 & 0 & -\frac{\sqrt{3}}{2} & -\frac{1}{2} & 0 & 0 & 0 & 0 & 0 & 0 & 0 \\
0 & 0 & \frac{1}{2} & 0 & -\frac{\sqrt{3}}{2} & 0 & 0 & 0 & 0 & 0 & 0 \\
0 & 1 & \frac{\sqrt{3}}{2} & 0 & -\frac{1}{2} & 0 & 0 & 0 & 0 & 0 & 0 \\
0 & 0 & 0 & 0 & \frac{\sqrt{3}}{2} & -1 & 0 & -\frac{\sqrt{3}}{2} & 0 & 0 & 0 \\
0 & 0 & 0 & 0 & \frac{1}{2} & 0 & 0 & \frac{1}{2} & 0 & 0 & 0 \\
0 & 0 & 0 & \frac{\sqrt{3}}{2} & 0 & 1 & -\frac{\sqrt{3}}{2} & 0 & 0 & 0 & 0 \\
0 & 0 & 0 & 0 & 0 & 0 & -\frac{1}{2} & 0 & \frac{\sqrt{3}}{2} & 0 & 1 \\
0 & 0 & 0 & 0 & 0 & 0 & 0 & \frac{\sqrt{3}}{2} & \frac{1}{2} & 0 & 0 \\
0 & 0 & 0 & 0 & 0 & 0 & 0 & -\frac{1}{2} & -\frac{\sqrt{3}}{2} & 1 & 0 \\
\hline
0 & 0 & 0 & \frac{1}{2} & 0 & 0 & \frac{1}{2} & 0 & 0 & 0 & 0 \\
0 & 0 & 0 & 0 & 0 & 0 & \frac{\sqrt{3}}{2} & 0 & -\frac{1}{2} & 0 & 0
\end{array}\right] .$$

$$\mathbf{A}_{21} \qquad\qquad \mathbf{A}_{22}$$

(3.2.25)

Vector **P** associated with matrix **A** given by equation (3.2.25) takes the form

$$\mathbf{P}_1 = \begin{bmatrix} P_1 \\ P_2 \\ P_3 \\ P_4 \\ P_5 \\ P_6 \\ P_7 \\ P_8 \\ P_9 \\ P_{10} \end{bmatrix}, \qquad \mathbf{P}_2[P_{11}] \ . \tag{3.2.26}$$

By using equation (3.2.21), **W** takes the form

$$\mathbf{W} = \begin{bmatrix} \frac{2\sqrt{3}}{3} & 0 & -\frac{2\sqrt{3}}{3} & 0 & -\frac{\sqrt{3}}{3} & -1 & +\frac{\sqrt{3}}{3} & 0 & 0 & 0 \\ -1 & 0 & -1 & 0 & -1 & 0 & -1 & 0 & -1 & 0 \end{bmatrix}. \tag{3.2.27}$$

Also in this case the elements of matrices $\mathbf{A}_{21}$ and $\mathbf{A}_{22}$ are dependent; they are linear combinations of the elements of the independent matrices $\mathbf{A}_{11}$ and $\mathbf{A}_{12}$ as given by equation (3.2.20).

The forces induced by the prestressing of the tensegric shell shown in Fig. 3.2.2(b) can be determined as a function of P_{11} by using the independent part of matrix **A** and equation (3.2.8). The fact that it is possible to determine the forces induced by the prestressing indicates that prestressing of this tensegric shell is feasible and requirement (i) is satisfied.

The cable net shown in Fig. 3.2.3 has the typical shape of the saddle roof shown in Fig. 1.1.5(b). In the actual structure there are many cables in the two directions. In this example there are two cables in the x direction and three cables in the y direction. This small number of elements was chosen to highlight the features of such a cable roof without clouding the analysis with excess mathematics. Each cable takes the shape of a parabola. The roof is composed of 17 elements with six free nodes. In this case $m = 18$ and $n = 17$; thus the cable net is an under-constrained cabled structure.

Matrix **A** is composed of 18 rows and 17 columns. By considering equilibrium in the x, y and z directions of modes 1, 2, 3, 4, 5 and 6, matrix **A** takes the form

$$\mathbf{A} = \begin{bmatrix}
0 & 0 & 0 & a & -b & 0 & 0 & 0 & 0 & 0 & 0 & 0 & 0 & 0 & 0 & 0 & 0 \\
-c & 0 & 0 & 0 & 0 & 0 & 0 & 1.0 & 0 & 0 & 0 & 0 & 0 & 0 & 0 & 0 & 0 \\
d & 0 & 0 & -e & f & 0 & 0 & 0 & 0 & 0 & 0 & 0 & 0 & 0 & 0 & 0 & 0 \\
0 & 0 & 0 & 0 & b & -b & 0 & 0 & 0 & 0 & 0 & 0 & 0 & 0 & 0 & 0 & 0 \\
0 & -c & 0 & 0 & 0 & 0 & 0 & 0 & 1.0 & 0 & 0 & 0 & 0 & 0 & 0 & 0 & 0 \\
0 & d & 0 & 0 & -f & -f & 0 & 0 & 0 & 0 & 0 & 0 & 0 & 0 & 0 & 0 & 0 \\
0 & 0 & 0 & 0 & 0 & b & -a & 0 & 0 & 0 & 0 & 0 & 0 & 0 & 0 & 0 & 0 \\
0 & 0 & -c & 0 & 0 & 0 & 0 & 0 & 0 & 1.0 & 0 & 0 & 0 & 0 & 0 & 0 & 0 \\
0 & 0 & d & 0 & 0 & f & -e & 0 & 0 & 0 & 0 & 0 & 0 & 0 & 0 & 0 & 0 \\
0 & 0 & 0 & 0 & 0 & 0 & 0 & 0 & 0 & 0 & a & -b & 0 & 0 & 0 & 0 & 0 \\
0 & 0 & 0 & 0 & 0 & 0 & 0 & -1.0 & 0 & 0 & 0 & 0 & 0 & 0 & c & 0 & 0 \\
0 & 0 & 0 & 0 & 0 & 0 & 0 & 0 & 0 & 0 & -e & f & 0 & 0 & d & 0 & 0 \\
0 & 0 & 0 & 0 & 0 & 0 & 0 & 0 & 0 & 0 & 0 & b & -b & 0 & 0 & 0 & 0 \\
0 & 0 & 0 & 0 & 0 & 0 & 0 & 0 & -1.0 & 0 & 0 & 0 & 0 & 0 & 0 & c & 0 \\
0 & 0 & 0 & 0 & 0 & 0 & 0 & 0 & 0 & 0 & 0 & -f & -f & 0 & 0 & d & 0 \\
0 & 0 & 0 & 0 & 0 & 0 & 0 & 0 & 0 & 0 & 0 & 0 & b & a & 0 & 0 & 0 \\
0 & 0 & 0 & 0 & 0 & 0 & 0 & 0 & 0 & -1.0 & 0 & 0 & 0 & 0 & 0 & 0 & c \\
0 & 0 & 0 & 0 & 0 & 0 & 0 & 0 & 0 & 0 & 0 & 0 & f & -e & 0 & 0 & d
\end{bmatrix} \tag{3.2.28}$$

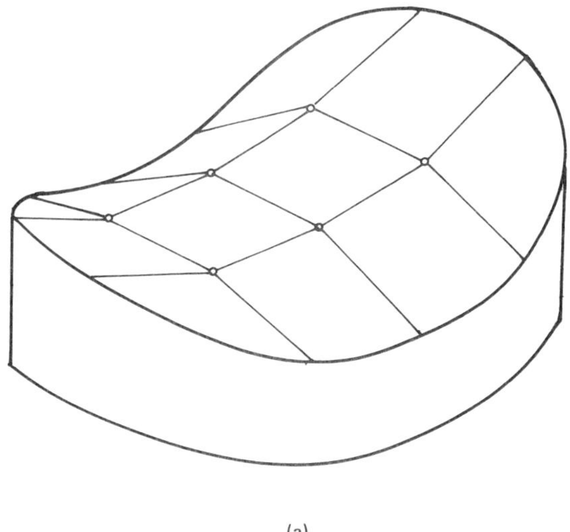

(a)

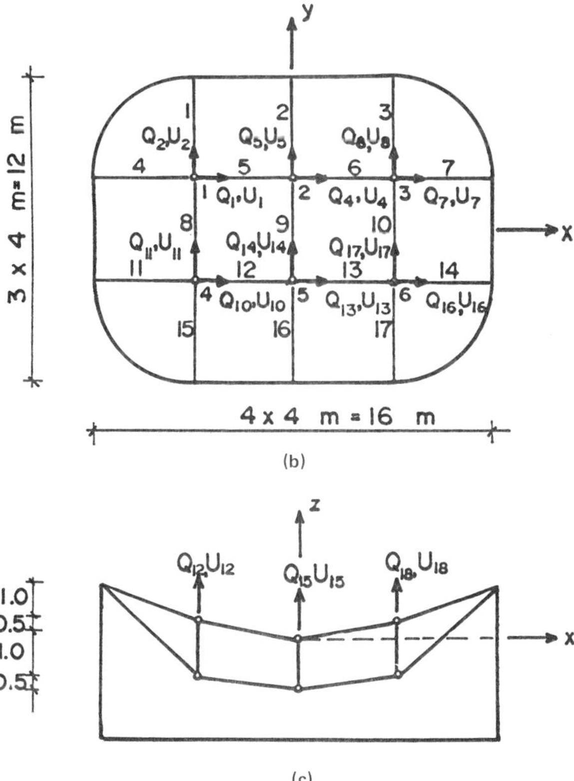

(c)

Fig. 3.2.3 — A typical cable net roof.

in which $a = 0.937$, $b = 0.993$, $c = 0.894$, $d = 0.447$, $e = 0.351$ and $f = 0.124$. By using Gaussian elimination it is found that two rows, rows 5 and 18 indicating equilibrium in the y direction at node 2 and in the z direction at node 6, are functions of the rest of the rows. Matrix **A** is rearranged by interchanging rows 5 and 17 and is partitioned accordingly; it takes the form

$$
\mathbf{A} = \left[\begin{array}{cccccccccccccccc|c}
0 & 0 & 0 & a & -b & 0 & 0 & 0 & 0 & 0 & 0 & 0 & 0 & 0 & 0 & 0 & 0 \\
-c & 0 & 0 & 0 & 0 & 0 & 0 & 1.0 & 0 & 0 & 0 & 0 & 0 & 0 & 0 & 0 & 0 \\
d & 0 & 0 & -e & f & 0 & 0 & 0 & 0 & 0 & 0 & 0 & 0 & 0 & 0 & 0 & 0 \\
0 & 0 & 0 & 0 & b & -b & 0 & 0 & 0 & 0 & 0 & 0 & 0 & 0 & 0 & 0 & 0 \\
0 & 0 & 0 & 0 & 0 & 0 & 0 & 0 & 0 & -1.0 & 0 & 0 & 0 & 0 & 0 & 0 & c \\
0 & d & 0 & 0 & -f & -f & 0 & 0 & 0 & 0 & 0 & 0 & 0 & 0 & 0 & 0 & 0 \\
0 & 0 & 0 & 0 & 0 & b & -a & 0 & 0 & 0 & 0 & 0 & 0 & 0 & 0 & 0 & 0 \\
0 & 0 & -c & 0 & 0 & 0 & 0 & 0 & 0 & 1.0 & 0 & 0 & 0 & 0 & 0 & 0 & 0 \\
0 & 0 & d & 0 & 0 & f & -e & 0 & 0 & 0 & 0 & 0 & 0 & 0 & 0 & 0 & 0 \\
0 & 0 & 0 & 0 & 0 & 0 & 0 & 0 & 0 & 0 & a & -b & 0 & 0 & 0 & 0 & 0 \\
0 & 0 & 0 & 0 & 0 & 0 & 0 & -1.0 & 0 & 0 & 0 & 0 & 0 & 0 & c & 0 & 0 \\
0 & 0 & 0 & 0 & 0 & 0 & 0 & 0 & 0 & 0 & -e & f & 0 & 0 & d & 0 & 0 \\
0 & 0 & 0 & 0 & 0 & 0 & 0 & 0 & 0 & 0 & 0 & b & -b & 0 & 0 & 0 & 0 \\
0 & 0 & 0 & 0 & 0 & 0 & 0 & 0 & -1.0 & 0 & 0 & 0 & 0 & 0 & 0 & c & 0 \\
0 & 0 & 0 & 0 & 0 & 0 & 0 & 0 & 0 & 0 & 0 & -f & -f & 0 & 0 & d & 0 \\
0 & 0 & 0 & 0 & 0 & 0 & 0 & 0 & 0 & 0 & 0 & 0 & b & a & 0 & 0 & 0 \\
\hline
0 & -c & 0 & 0 & 0 & 0 & 0 & 0 & 1.0 & 0 & 0 & 0 & 0 & 0 & 0 & 0 & 0 \\
0 & 0 & 0 & 0 & 0 & 0 & 0 & 0 & 0 & 0 & 0 & 0 & f & -e & 0 & 0 & d
\end{array}\right] .
\tag{3.2.29}
$$

$\mathbf{A}_{11}$ is a 16×16 matrix with a determinant different from zero. $\mathbf{A}_{12}$ is a 16×1 matrix, $\mathbf{A}_{21}$ is a 2×16 matrix and $\mathbf{A}_{22}$ is a 2×1 matrix. The elements of matrices $\mathbf{A}_{21}$ and $\mathbf{A}_{22}$ are a function of the elements of matrices $\mathbf{A}_{11}$ and $\mathbf{A}_{12}$. They are to be found by using equation (3.2.20) in which **W** takes the form

$$
\mathbf{W} = \left[\begin{array}{cccccccc}
0.749 & 1.000 & 2.000 & 0.249 & 0.000 & -2.000 & 0.000 & 0.000 \\
-0.375 & -0.500 & -1.000 & -0.249 & 0.500 & 0.000 & -0.375 & 0.5000
\end{array}\right.
$$

$$
\left.\begin{array}{cccccccc}
0.000 & -0.749 & 1.000 & -2.000 & -0.249 & -1.000 & 2.000 & 0.000 \\
1.000 & 0.375 & -0.500 & 1.000 & 0.249 & 0.000 & 0.000 & 0.375
\end{array}\right] .
\tag{3.2.30}
$$

Vector **P** associated with matrix **A** given by equation (3.2.29) takes the form

$$\mathbf{P}_1 = \begin{bmatrix} P_1 \\ P_2 \\ P_3 \\ P_4 \\ P_5 \\ P_6 \\ P_7 \\ P_8 \\ P_9 \\ P_{10} \\ P_{11} \\ P_{12} \\ P_{13} \\ P_{14} \\ P_{15} \\ P_{16} \end{bmatrix}, \qquad \mathbf{P}_2 = [P_{17}] \ . \tag{3.2.31}$$

By using equation (3.2.8) the forces induced by prestressing given by $\mathbf{P}_1$ can be found as a function of P_{17}, the force induced in member 17. The fact that it is possible to determine the forces induced by the prestressing indicates that prestressing of this cable net is feasible and requirement (i) is satisfied.

Another type of saddle roof is shown in Fig. 3.2.4. The net is composed of 14 cables.

The surface of the net takes the form of a hyperbolic paraboloid and the nodes of the net fit

$$z = 2 \times 10^{-2} \, (x^2 - y^2) \ . \tag{3.2.32}$$

The net is composed of 64 members connected at 25 different free nodes. Matrix **A** is composed of 75 rows and 64 columns and the net is an under-constrained cabled structure. In the case where prestressing is symmetrical, as shown in Fig. 3.2.5, 20 different forces are induced in the members of the net by prestressing. There are 30 equilibrium conditions to satisfy at the ten different nodes. Because equilibrium in the x direction is satisfied automatically at nodes 1, 2, 3 and 7 and in the y direction at nodes 7, 8, 9 and 10, matrix **A** is composed of 22 rows and 21 columns. The first and second rows are associated with the equilibrium at node 1 in the y and z directions. The third and fourth rows are associated with the equilibrium at node 2 in the y and z directions and so on until the twenty-first and twenty-second rows are associated with the equilibrium at node 10 in the x and z directions:

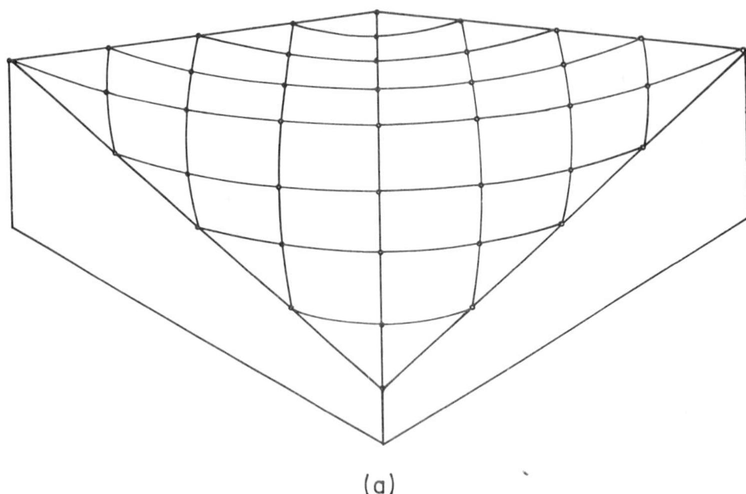

(a)

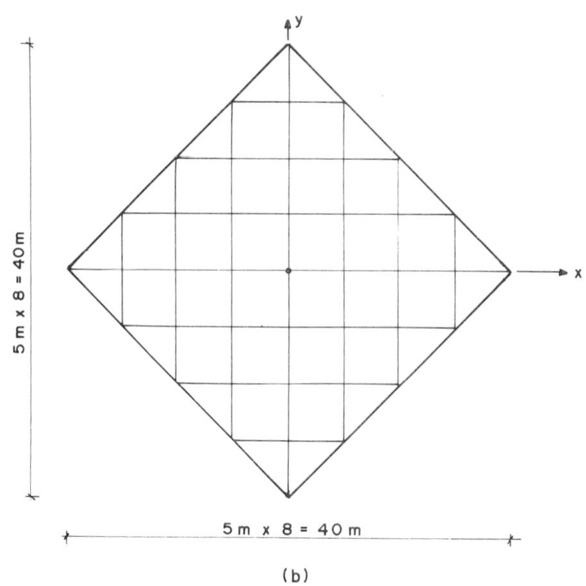

(b)

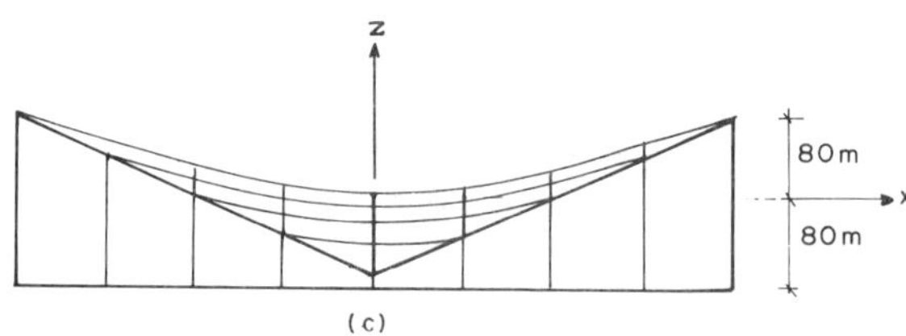

(c)

Fig. 3.2.4 — A cable net in the shape of a hyperbolic paraboloid.

$$
\mathbf{A}=\begin{bmatrix}
0&0&0&0&0&0&0&0&0&0&0&0&0&0&0&a&0&0&0&b\\
0&0&0&0&0&0&0&0&0&0&0&0&0&0&0&c&d&0&0&-e\\
0&0&0&0&0&0&0&f&0&0&0&0&0&0&0&-a&0&0&0&0\\
0&0&0&0&0&0&0&g&0&0&0&d&0&0&0&-c&0&0&0&0\\
0&0&0&0&0&0&0&0&0&0&-f&h&0&0&0&0&0&0&0&0\\
0&0&0&0&0&0&f&0&0&0&0&0&0&0&-a&0&0&0&0&0\\
0&0&0&0&0&0&g&0&0&0&g&-i&0&0&-c&0&0&0&0&0\\
h&0&0&0&0&0&0&-f&0&0&0&0&0&0&0&0&0&0&0&0\\
i&0&0&d&0&0&0&-g&0&0&0&0&0&0&0&0&0&0&0&0\\
0&0&0&h&0&-f&0&0&0&0&0&0&0&0&0&0&0&0&0&0\\
0&0&h&0&0&0&-f&0&0&0&0&0&0&0&0&0&0&0&0&0\\
0&0&i&i&0&g&-g&0&0&0&0&0&0&0&0&0&0&0&0&0\\
0&0&0&0&0&f&0&0&0&0&0&0&0&-a&0&0&0&0&0&0\\
0&0&0&0&0&0&0&0&h&-f&0&0&0&0&0&0&0&0&0&0\\
0&0&0&0&0&-g&0&0&i&-g&0&0&0&0&0&0&0&0&0&0\\
-d&d&0&0&0&0&0&0&0&0&0&0&0&0&0&0&0&0&0&0\\
0&h&0&0&-f&0&0&0&0&0&0&0&0&0&0&0&0&0&0&0\\
0&-i&d&0&g&0&0&0&0&0&0&0&0&0&0&0&0&0&0&0\\
0&0&0&0&f&0&0&0&0&0&0&0&-a&0&0&0&0&0&0&0\\
0&0&0&0&-g&0&0&0&-d&0&0&0&0&0&0&0&0&0&0&0\\
0&0&0&0&0&0&0&0&0&0&0&0&a&0&0&0&0&0&-b&0\\
0&0&0&0&0&0&0&0&0&0&0&0&-c&0&0&0&0&d&e&0
\end{bmatrix}
\tag{3.2.33}
$$

in which $a=0.894$, $b=0.819$, $c=0.447$, $d=0.198$, $e=0.573$, $f=0.958$, $g=0.287$, $h=0.995$ and $i=0.099$.

By using Gaussian elimination it is found that three rows, 8, 12 and 18, which are associated with the equilibrium conditions in the z direction at nodes 4, 5 and 8, are a function of the rest of the rows. This fact indicates that the rank of **A** is 19 only. The forces induced in the members by prestressing can be found as a function of P_{20} by using the independent part of matrix **A** which is composed of the appropriate matrices: $\mathbf{A}_{11}$ (19×19) and matrix $\mathbf{A}_{12}$ (19×1). The fact that it is possible to determine the forces induced by prestressing indicates that prestressing of this cable net is feasible and requirement (i) is satisfied.

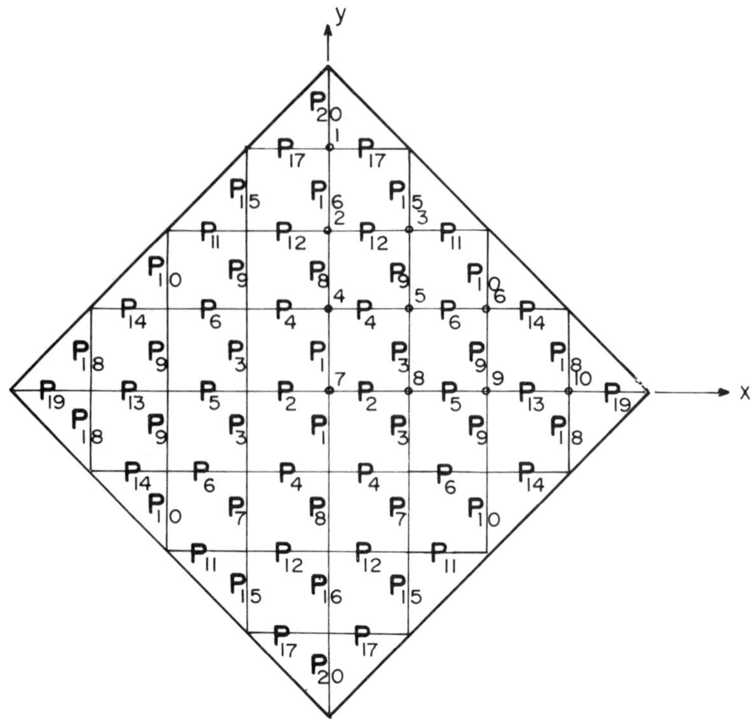

Fig. 3.2.5 — Prestressing of the net shown in Fig. 3.2.3.

These examples illustrate that an under-constrained cabled structure is prestressable in the cases where equation (3.2.8) can be used to determine the accurate forces induced by prestressing into the members of the structure. In under-constrained cabled structures this equation can be used only where matrix $\mathbf{A}$ can be partitioned into an independent part composed of $\mathbf{A}_{11}$ and $\mathbf{A}_{12}$ and a dependent part composed of $\mathbf{A}_{21}$ and $\mathbf{A}_{22}$. Matrix $\mathbf{A}$ can be partitioned in such a way, only where its rank is less than n. Matrix $\mathbf{A}$ has a rank less than n where the under-constrained cabled structure has a specific configuration. Most structures which are not indeterminate have configurations which cannot be prestressed. For example, the determinate structure shown in Fig. 2.5.1 cannot be prestressed and none of its members can be a cable. Gaussian elimination of matrix $\mathbf{A}$ of this structure, given by equation (2.5.4), takes the form

$$\mathbf{A} = \begin{bmatrix} \frac{1}{2} & 0 & 0 & 0 \\ 0 & \dfrac{\sqrt{3}}{2} & 1 & 0 \\ 0 & 0 & -1 & 0 \\ 0 & 0 & 0 & \dfrac{\sqrt{3}}{2} \end{bmatrix} . \tag{3.2.34}$$

The fact that no row is composed of elements all equal to zero indicates that the rank of matrix **A** is 4, equal to n. In this case, matrix **A** cannot be partitioned into $\mathbf{A}_{11}$, $\mathbf{A}_{12}$, $\mathbf{A}_{21}$ and $\mathbf{A}_{22}$ in the method discussed before, where $\mathbf{A}_{11}$ is a square matrix with a determinant different from zero and where the elements of $\mathbf{A}_{21}$ and $\mathbf{A}_{22}$ are functions of the elements of $\mathbf{A}_{11}$ and $\mathbf{A}_{12}$. Equation (3.2.8) cannot be used to determine the forces induced in the elements of the structure by prestressing, and the structure is not prestressable. It can be seen that by using matrix **A** given by equation (3.2.34), equation (3.2.1) of the prestressing forces of this structure takes the form

$$
\begin{aligned}
&\tfrac{1}{2}P_1 = 0\\
&\frac{\sqrt{3}}{2}P_2 + P_3 = 0\\
&\qquad -P_3 = 0\\
&\qquad \frac{\sqrt{3}}{2}P_4 = 0\ .
\end{aligned}
\tag{3.2.35}
$$

There is only one solution to this set: $P_1 = P_2 = P_3 = P_4 = 0$. The structure is not a prestressable structure, requirement (i) cannot be satisfied and it is not a feasible structure when some of its elements are made of cables.

Also the prestressing of the structure shown in Fig. 2.5.2 is not feasible and no under-constrained cabled structure can take this configuration. Gaussian elimination of matrix **A** of this structure, given by equation (2.5.6), takes the form

$$
\mathbf{A} = \begin{bmatrix} \frac{1}{2} & -1 & 0 \\ 0 & 1 & -\frac{1}{2} \\ 0 & 0 & \frac{\sqrt{3}}{2} \\ 0 & 0 & 0 \end{bmatrix} . \tag{3.2.36}
$$

The elements of one row only are equal to zero which implies that the rank of matrix **A** is 3 equal to n. Also in this case it is impossible to partition matrix **A** into $\mathbf{A}_{11}$, $\mathbf{A}_{12}$, $\mathbf{A}_{21}$ and $\mathbf{A}_{22}$ according to the method discussed before. This indicates that equation (3.2.8) cannot be used to determine the forces induced in the members of the structure by prestressing. Thus, prestressing of the structure is not feasible and requirement (i) cannot be satisfied.

3.3 THE FORCES INDUCED BY PRESTRESSING

The actual prestressing of a cabled structure takes place by shortening some of the cables or in the case of tensegric shells also by lengthening some of the bars. The prestressing is to the level required to ensure a minimum amount of tension in all cables under all possible loads. The magnitude of the forces induced by the prestressing can be determined by using equation (3.2.1). For a fully constrained cabled structure the number m of equations is less than the number n of the unknown internal forces. In this case, $n-m$ indicates the degrees of freedom of prestressing. $n-m$ is the number of members that can be prestressed independently to a required magnitude. The forces induced in the rest of the members can be found by using equation (3.2.1). In the case of the structure shown in Fig. 3.2.2(a), equation (3.2.1) may be written as given by equation (3.2.8). It can be seen that the net has one degree of freedom of prestressing and it can be prestressed by prestressing member 5 only. The forces induced in members 1, 2, 3 and 4 can be determined by using equation (3.2.8). In the case where member 5 is prestressed to 5 kN, the forces induced in the rest of the members by prestressing are found by using equation (3.2.8) and are shown in Fig. 3.3.1.

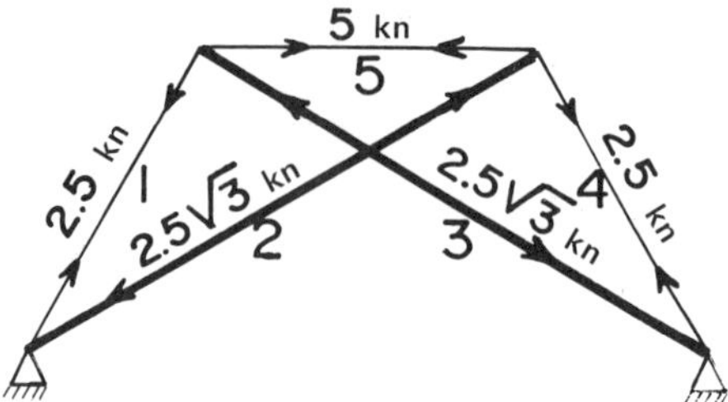

Fig. 3.3.1 — The prestressing of the tensegric shell shown in Fig. 3.2.2(a).

The arrows on the members in Fig. 3.3.1 indicate the forces applied by the member to the node. In the case of a member under compression the arrows take the form as bars 2 and 3 in Fig. 3.3.1. In the case of a member under tension the arrows take the form $\rightarrow - \leftarrow$ as cables 1, 5 and 4 in Fig. 3.3.1.

For the structure shown in Fig. 3.2.1(a), matrices $\mathbf{A}_{11}$, $\mathbf{A}_{12}$, $\mathbf{P}_1$ and $\mathbf{P}_2$ take the forms given by equations (3.2.10) and (3.2.11). In this case the net has two degrees of freedom of prestressing; it can be prestressed by inducing prestressing forces in members 3 and 4. The force induced in member 3 is independent of that induced in member 4. For example, cables 3 and 4 can be prestressed to 5 kN each and the forces induced by prestressing can be analysed by using equation (3.2.8). They are shown in Fig. 3.3.2(a). The case where cables 3 and 4 are prestressed to 2.5 kN and 5 kN, respectively, is shown in Fig. 3.3.2(b).

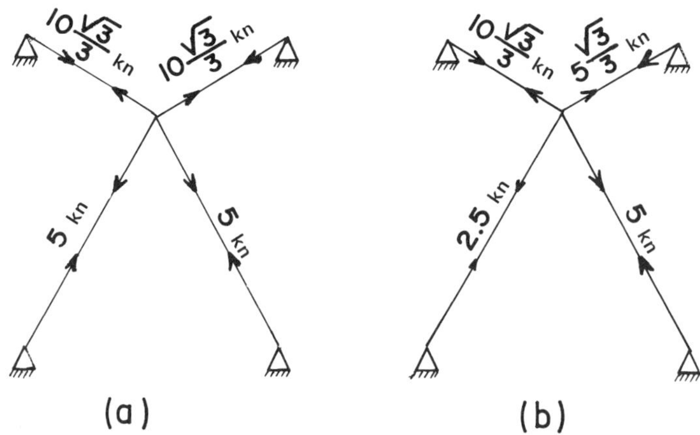

Fig. 3.3.2 — The prestressing of the cable net shown in Fig. 3.2.1(b).

Also, for an under-constrained cabled structure, equation (3.2.8) can be used to determine the forces induced by the prestressing. For example, for the cable net shown in Fig. 3.2.1(b) the forces induced in all cables by prestressing can be found as a function of the force induced in cable 8. By using $\mathbf{A}_{11}$, $\mathbf{A}_{12}$, $\mathbf{P}_1$ and $\mathbf{P}_2$ given by equations (3.2.18) and (3.2.19), the forces induced in the rest of the cables by prestressing, where $\mathbf{P}_1$ is prestressed to 5 kN, can be determined by using equation (3.2.8) and is shown in Fig. 3.3.3.

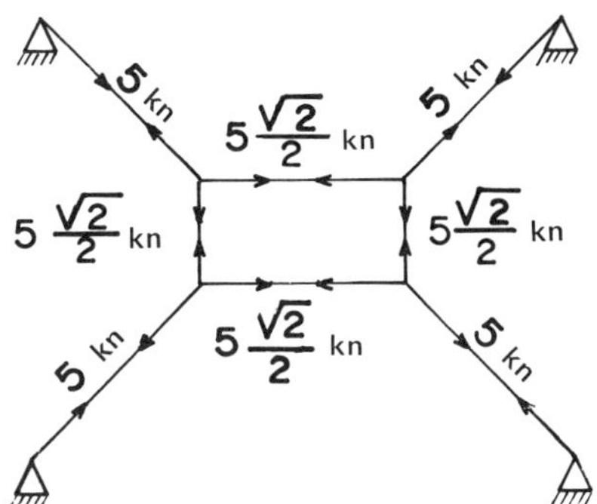

Fig. 3.3.3 — The forces induced by prestressing of the cable net shown in Fig. 3.2.1(b).

For the tensegric shell shown in Fig. 3.2.2(b) the forces induced by prestressing can be found by using matrices $\mathbf{A}_{11}$ and $\mathbf{A}_{12}$ given by equation (3.2.25) and by using equation (3.2.8). The case where bar 11 is prestressed to 5 kN is shown in Fig. 3.3.4.

For the cable net shown in Fig. 3.2.3 the forces induced by prestressing can be found by using equation (3.2.8) considering matrices $\mathbf{A}_{11}$ and $\mathbf{A}_{12}$ given by equation (3.2.29). The case where a tension of 20.0 kN is induced in element 17 is shown in Fig. 3.3.5.

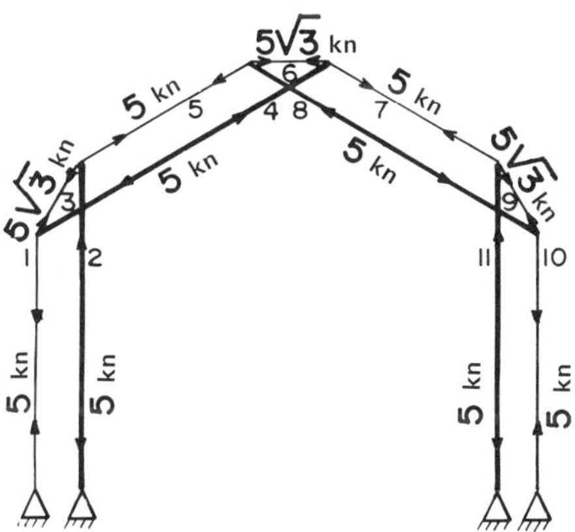

Fig. 3.3.4 — The forces induced by prestressing of the tensegric shell shown in Fig. 3.2.2(b).

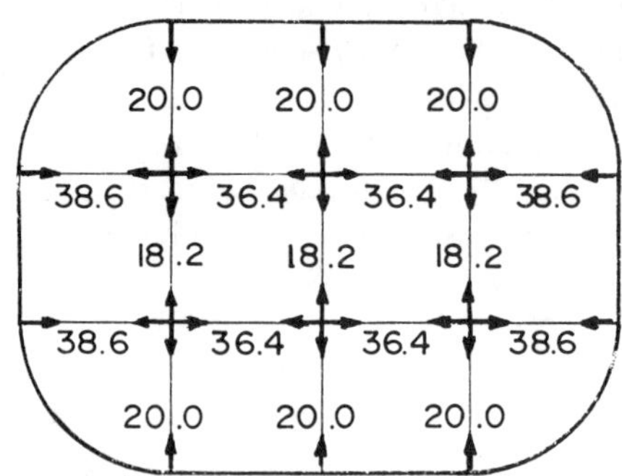

Fig. 3.3.5 — The tension in kilonewtons induced in the elements by prestressing.

The fact that $\mathbf{A}_{12}$ is a 16×1 matrix indicates that there is one degree of freedom in prestressing only. It is enough to induce the required force by prestressing member 17 and the forces induced in the rest of the members are uniquely defined by using equation (3.2.8).

For the cable net shown in Fig. 3.2.4 the forces induced by prestressing can be found by using equation (3.2.8) considering the appropriate matrices $\mathbf{A}_{11}$ and $\mathbf{A}_{12}$. The case where a tension of 100 kN is induced in element 20 by prestressing is shown in Fig. 3.3.6.

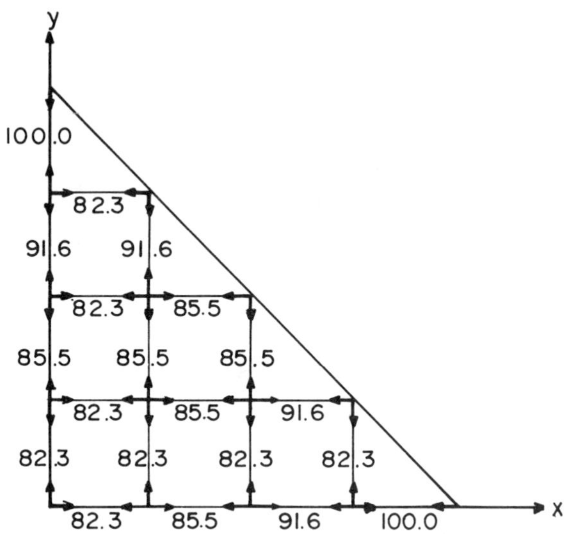

Fig. 3.3.6 — The tension in kilonewtons induced in the elements by prestressing.

Also in this case the fact that $\mathbf{A}_{12}$ is a 19×1 matrix indicates that there is one degree of freedom of prestressing only. It is enough to induce the required force by prestressing one member only and the forces shown in Fig. 3.3.5 are induced in the rest of the members.

The fact that in the analysis the force in one member was assumed in order to predict the forces induced in the rest of the members does not imply that this member should be prestressed. Prestressing of any one of the members to the required level will induce the prestressing forces predicted by the analysis. In the case where the degree of freedom of the prestressing is larger than 1, prestressing of members equal to the degrees of freedom of prestressing to the required level is enough to ensure that the analysed forces are induced in the members of the structure.

In order to satisfy requirement (ii) of a satisfactory cabled structure discussed in section 3.1, it is important to ensure that tension is induced in the cables of the structure by prestressing. For example, members 2 and 3 in the tensegric shell shown in Fig. 3.3.1 can be bars only, because prestressing of the structure induces compression in them. It can be seen that the nets shown in Fig. 3.3.2 and 3.3.3 are satisfactory. The tensegric shell shown in Fig. 3.3.4 is satisfactory where members 2, 4, 8 and 11 are bars. Otherwise this structure is not a satisfactory cabled structure.

The magnitude of the prestressing is determined to ensure a minimum amount of tension in all cables under all expected loads.

3.4 THE DEFORMATION DUE TO PRESTRESSING

The deformation of the cabled structure because of the prestressing is due to the elasticity of the members and is small. In most practical cases the distortion of the structure caused by this deformation is small and can be ignored.

In the design of a cabled structure, two methods can be used considering the distortion of the structure because of the prestressing of its members. One possible method is to design the length of the members so that, where the structure is prestressed, it gets its given shape. Another method is to find the distortion of a given configuration of the structure caused by the prestressing.

For example, for the cable net shown in Fig. 3.2.1(a), considering the forces induced by prestressing shown in Fig. 3.3.2(a), the members should be of the following lengths where the first method is followed:

$$
\begin{aligned}
l_1 &= a\,\frac{2\sqrt{3}}{3}\left(1-\frac{10\sqrt{3}}{3E,A}\right)\\
l_2 &= a\,\frac{2\sqrt{3}}{2}\left(1-\frac{10\sqrt{3}}{3E_2A_2}\right)\\
l_3 &= a2\left(1-\frac{5}{E_3A_3}\right)\\
l_4 &= a2\left(1-\frac{5}{E_4A_4}\right)
\end{aligned}
\qquad (3.4.1)
$$

where E_i and A_i are the modules of elasticity and the area of the cross-section of member i. In this casse, because the prestressing forces are given in kilonewtons, the units of E_i and A_i should be in kilonewtons and the unit of length is that used to determine a. The configuration of this prestressed cabled structure takes the exact shape shown in Fig. 3.2.1(a).

For a fully constrained cabled structure where the second method is followed, equation (2.5.1) can be used to determine the distortion of the configuration of the structure. The change in the length of the members of the tensergic shell shown in Fig. 3.2.2.(a), where prestressed to the level shown in Fig. 3.3.1 by lengthening member 2, takes the form

$$\begin{aligned}
\Delta_1 &= \frac{5}{E_1 A_1} b \\
\Delta_3 &= \frac{-5}{E_3 A_3} (a+b) \\
\Delta_4 &= \frac{5}{E_4 A_4} b \\
\Delta_5 &= \frac{5}{E_5 A_5} a \ .
\end{aligned} \tag{3.4.2}$$

Also in this case, E_i, A_i should be in kilonewtons and the unit of length is that used to determine a and b. The change in length of member 2 depends on its lengthening during the prestressing which is determined later.

The nodal displacements can be analysed by equation (2.2.9):

$$\begin{aligned}
\Delta_1 &= \tfrac{1}{2} U_1 + \frac{\sqrt{3}}{2} U_2 \\
0 &= -\Delta_2 + \frac{\sqrt{3}}{2} U_3 + \tfrac{1}{2} U_4 \\
\Delta_3 &= \frac{\sqrt{3}}{2} U_1 + \tfrac{1}{2} U_2 \\
\Delta_4 &= -\tfrac{1}{2} U_3 + \frac{\sqrt{3}}{2} U_4 \\
\Delta_5 &= -U_1 + U_3 \ .
\end{aligned} \tag{3.4.3}$$

Equation (3.4.3) is a set of five equations with five unknowns: U_1, U_2, U_3, U_4 and Δ_2. The actual change in the length of member 2 denoted by $\overline{\Delta}_2$ is composed of two components:

(a) the change in length due to the nodal displacements given by Δ_2;
(b) the change in length due to the force induced into it by prestressing.

By considering these two aspects $\bar{\Delta}_2$ takes the form

$$\bar{\Delta}_2 = \Delta_2 + \frac{5(a+b)}{E_2A_2} \ . \tag{3.4.4}$$

The change in length of cables 1 and 2 of the fully constrained cable net shown in Fig. 3.2.1(a), due to prestressing, can be analysed by the same method. When prestressing is to the level shown in Fig. 3.3.2(b), due to the shortening of cables 3 and 4, the changes in length of members 1 and 2 take the forms

$$\begin{aligned} \Delta_1 &= \frac{10}{E_1A_1} a \\ \Delta_2 &= \frac{5}{E_2A_2} a \ . \end{aligned} \tag{3.4.5}$$

By using equation (2.2.9), the nodal displacements U_1 and U_2 take the forms

$$\begin{aligned} \Delta_1 &= \frac{\sqrt{3}}{2} U_1 - \tfrac{1}{2} U_2 \\ \Delta_2 &= -\frac{\sqrt{3}}{2} U_1 - \tfrac{1}{2} U_2 \\ 0 &= -\Delta_3 + \tfrac{1}{2}U_1 - \frac{\sqrt{3}}{2} U_2 \\ 0 &= -\Delta_4 - \tfrac{1}{2}U_1 + \frac{\sqrt{3}}{2} U_2 \ . \end{aligned} \tag{3.4.6}$$

Equation (3.4.6) is a set of four equations. By using it, U_1, U_2, Δ_3 and Δ_4 can be determined. The actual shortening of the length of members 3 and 4 required to prestress the cable net is

$$\begin{aligned} \bar{\Delta}_3 &= \Delta_3 + \frac{5}{E_3A_3} a \\ \bar{\Delta}_4 &= \Delta_4 + \frac{10}{E_4A_4} a \ . \end{aligned} \tag{3.4.7}$$

The under-constrained cabled structure is discussed by considering the case of the cable net shown in Fig. 3.2.1(b) first. The case where it is prestressed by shortening cable 8 is investigated. The change in length of cables 1, 2, 3, 4, 5, 6 and 7 can be determined by using equation (2.4.3) which takes the form

$$\begin{aligned}
\Delta_1 &= \frac{5\sqrt{2}}{2E_1A_1}\,a \\
\Delta_2 &= \frac{5\sqrt{2}}{2E_2A_2}\,c \\
\Delta_3 &= \frac{5\sqrt{2}}{2E_3A_3}\,a \\
\Delta_4 &= \frac{5\sqrt{2}}{2E_4A_4}\,c \\
\Delta_5 &= \frac{5\sqrt{2}}{2E_5A_5}\,c \\
\Delta_6 &= \frac{5\sqrt{2}}{2E_5A_5}\,a \\
\Delta_7 &= \frac{5\sqrt{2}}{2E_7A_7}\,b \;.
\end{aligned} \tag{3.4.8}$$

Δ_8 depends on the shortening of member 8 during prestressing and cannot be determined by equation (2.4.3). The relationship between the nodal displacements and the changes in the lengths of the members is found by using equation (2.3.8) and matrix **A** given by equation (3.2.14):

$$\begin{aligned}
\Delta_1 &= \frac{\sqrt{2}}{2}\,U_1 - \frac{\sqrt{2}}{2}\,U_2 \\
\Delta_2 &= -\,U_1 + U_3 \\
\Delta_3 &= -\frac{\sqrt{2}}{2}\,U_3 - \frac{\sqrt{2}}{2}\,U_4 \\
\Delta_4 &= U_2 - U_6 \\
\Delta_5 &= U_4 - U_8 \\
\Delta_6 &= \frac{\sqrt{2}}{2}\,U_5 - \frac{\sqrt{2}}{2}\,U_6 \\
\Delta_7 &= -\,U_5 + U_7 \\
0 &= \Delta_8 - \frac{\sqrt{2}}{2}\,U_7 + \frac{\sqrt{2}}{2}\,U_8 \;.
\end{aligned} \tag{3.4.9}$$

Equation (3.4.9) is a set of eight equations with nine unknowns U_1 to U_8 and Δ_8. Because there are more unknowns than equations in equation (3.4.9), it seems that in the prestressing of an under-constrained cabled structure it is impossible to determine the nodal displacements uniquely. The problem is investigated by examining equation (3.4.9). By using equation (2.3.8) and matrices $\mathbf{A}_{11}$, $\mathbf{A}_{12}$, $\mathbf{A}_{21}$ and $\mathbf{A}_{22}$, equation (3.4.9) takes the following form:

$$\mathbf{\Delta} = \begin{bmatrix} \mathbf{A}_{11}^{\mathrm{T}} & \mathbf{A}_{21}^{\mathrm{T}} \\ \mathbf{A}_{12}^{\mathrm{T}} & \mathbf{A}_{22}^{\mathrm{T}} \end{bmatrix} \mathbf{U} \ . \tag{3.4.10}$$

Vectors $\mathbf{\Delta}$ and $\mathbf{U}$ are partitioned into $\mathbf{\Delta}_1$, $\mathbf{\Delta}_2$, $\mathbf{U}_1$ and $\mathbf{U}_2$ in accordance with $\mathbf{A}_{11}^{\mathrm{T}}$, $\mathbf{A}_{21}^{\mathrm{T}}$, $\mathbf{A}_{12}^{\mathrm{T}}$ and $\mathbf{A}_{22}^{\mathrm{T}}$. $\mathbf{\Delta}_1$ and $\mathbf{\Delta}_2$ are associated with the change in length of the members in which $\mathbf{P}_1$ and $\mathbf{P}_2$ indicate the forces induced in them by the prestressing. $\mathbf{U}_1$ is associated with the equilibrium conditions which are determined by using matrices $\mathbf{A}_{11}$ and $\mathbf{A}_{12}$. $\mathbf{U}_2$ is associated with the equilibrium conditions which are determined by using matrices $\mathbf{A}_{21}$ and $\mathbf{A}_{22}$. By using $\mathbf{\Delta}_1$, $\mathbf{\Delta}_2$, $\mathbf{U}_1$ and $\mathbf{U}_2$, equation (3.4.10) takes the form

$$\begin{aligned} \mathbf{\Delta}_1 &= \mathbf{A}_{11}^{\mathrm{T}}\,\mathbf{U}_1 + \mathbf{A}_{21}^{\mathrm{T}}\,\mathbf{U}_2 \\ \mathbf{\Delta}_2 &= \mathbf{A}_{12}^{\mathrm{T}}\,\mathbf{U}_1 + \mathbf{A}_{22}^{\mathrm{T}}\,\mathbf{U}_2 \ . \end{aligned} \tag{3.4.11}$$

By using equation (3.2.20), equation (3.4.11) takes the form

$$\begin{aligned} \mathbf{\Delta}_1 &= \mathbf{A}_{11}^{\mathrm{T}}(\mathbf{U}_1 + \mathbf{W}^{\mathrm{T}}\mathbf{U}_2) \\ \mathbf{\Delta}_2 &= \mathbf{A}_{12}^{\mathrm{T}}(\mathbf{U}_1 + \mathbf{W}^{\mathrm{T}}\mathbf{U}_2) \ . \end{aligned} \tag{3.4.12}$$

Equation (3.4.12) indicates that not all the elements of vectors $\mathbf{U}_1$ and $\mathbf{U}_2$ are independent in equation (3.4.12). The independent elements can be defined by using vector $\mathbf{e}$ which takes the form

$$\mathbf{e} = \mathbf{U}_1 + \mathbf{W}^{\mathrm{T}}\mathbf{U}_2 \ . \tag{3.4.13}$$

In the case given by equation (3.4.9), $\mathbf{U}_1$ is associated with the equilibrium equations in the direction of U_1, U_2, U_3, U_4, U_5, U_6 and U_7. $\mathbf{U}_2$ is associated with the equilibrium equation in the U_8 direction. In this case, $\mathbf{U}_1$ and $\mathbf{U}_2$ take the forms

$$\mathbf{U} = \begin{bmatrix} \mathbf{U}_1 \\ \mathbf{U}_2 \end{bmatrix} , \quad \mathbf{U}_1 = \begin{bmatrix} U_1 \\ U_2 \\ U_3 \\ U_4 \\ U_5 \\ U_6 \\ U_7 \end{bmatrix} , \quad \mathbf{U}_2 = [U_8] \ . \tag{3.4.14}$$

By using **W** given by equation (3.2.22), vector **e** takes the form

$$\begin{aligned} e_1 &= U_1 + U_8 \\ e_2 &= U_2 + U_8 \\ e_3 &= U_3 + U_8 \\ e_4 &= U_4 - U_8 \\ e_5 &= U_5 - U_8 \\ e_6 &= U_6 + U_8 \\ e_7 &= U_7 - U_8 \ . \end{aligned} \tag{3.4.15}$$

By using equations (3.4.15) and (3.4.12), equation (3.4.9) takes the form

$$\begin{aligned} \Delta_1 &= \frac{\sqrt{2}}{2} l_1 - \frac{\sqrt{2}}{2} l_2 \\ \Delta_2 &= -l_1 + l_3 \\ \Delta_3 &= -\frac{\sqrt{2}}{2} l_3 - \frac{\sqrt{2}}{2} l_4 \\ \Delta_4 &= -l_4 - l_6 \\ \Delta_5 &= l_4 \\ \Delta_6 &= \frac{\sqrt{2}}{2} l_5 + \frac{\sqrt{2}}{2} l_6 \\ \Delta_7 &= -l_5 + l_7 \\ 0 &= -\Delta_8 - \frac{\sqrt{2}}{2} l_7 \ . \end{aligned} \tag{3.4.16}$$

Equation (3.4.16) is a set of eight equations with eight unknowns $\Delta_8, l_1, l_2, l_3, l_4, l_5, l_6$ and l_7. The actual shortening of the length of cable 8 takes the form

$$\bar{\Delta}_8 = \Delta_8 + \frac{5\sqrt{2}}{E_8 A_8} a \ . \tag{3.4.17}$$

By using equation (3.4.16), only vector **e** can be determined; it is impossible to determine explicitly the nodal displacements because of the prestressing of the structure, the physical meaning of which is discussed in the next chapter.

In order to determine the nodal displacements caused by the prestressing, the **geometrical non-linear effect** should be considered. This is the effect of the change in the configuration of the structure due to the nodal displacements. In most conventional structures the effect is small and can be ignored. In under-constrained cabled structures the effect is significant enough to determine their behaviour. The nodal displacements of under-constrained cabled structures caused by prestressing are studied by investigating the equilibrium conditions of statics at the nodes considering prestressing forces, and the deformed configuration of the structure. In this case, equation (3.2.1) takes the form

$$\mathbf{GP} = \mathbf{0} \ . \tag{3.4.18}$$

The elements of **G** are determined by considering the nodal displacements. The effect of the nodal displacements on the elements of matrix **G** can be studied by investigating the change in the geometrical configuration of the members of the structure.

The change in the configuration of member i takes the form shown in Fig. 3.4.1.

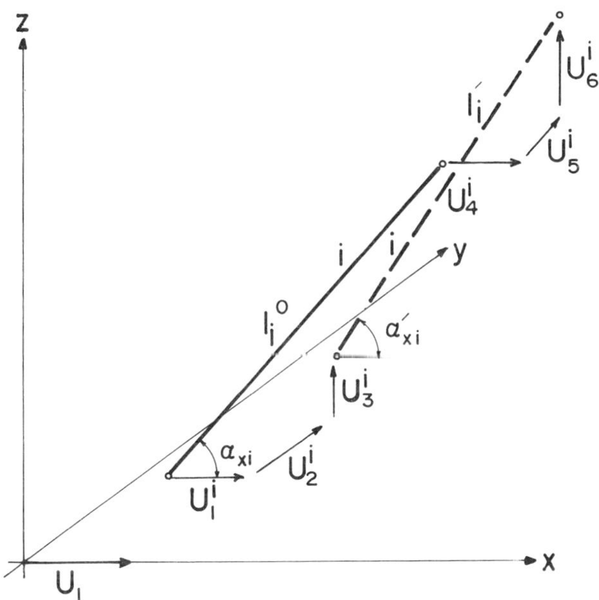

Fig. 3.4.1 — The displacement of member i.

Because of the nodal displacements the inclination α_{ix} changes to α_{ix}':

$$\cos \alpha_{ix}' = \frac{l_i^0 \cos \alpha_{ix} + (U_4^i - U_1^i)}{l_i'} \ . \tag{3.4.19}$$

l_i^0 is the original length of member i before the displacements of the nodes:

$$l_i^0 = (l_i^0 \cos \alpha_{ix})^2 + (l_i \cos \alpha_{iy})^2 + (l_i \cos \alpha_{iz})^2 \ . \tag{3.4.20}$$

l_i' is the length of the member considering the displacement of the nodes:

$$l_i' = \sqrt{[l_i^0 \cos \alpha_{ix} + (U_4^i - U_1^i)]^2 + [l_i \cos \alpha_{iy} + (U_5^i - U_2^i)]^2 + [l_i^0 \cos \alpha_{iz} + (U_6^i - U_5^i)]^2} \ . \tag{3.4.21}$$

For a small strain, i.e.

$$\left|\frac{U_4^i - U_1^i}{l_i^0}\right| \approx \left|\frac{U_5^i - U_2^i}{l_i^0}\right| \approx \left|\frac{U_6^i - U_5^i}{l_i^0}\right| \ll 1\ , \tag{3.4.22}$$

the second order of the strain can be ignored in equation (3.4.21) which takes the form

$$l_i' = l_i^0\sqrt{1+2\varepsilon}\ . \tag{3.4.23}$$

ε takes the form

$$\varepsilon = (\cos\alpha_{ix})\frac{(U_4^i - U_1^i)}{l_i^0} + (\cos\alpha_{iy})\frac{(U_5^i - U_2^i)}{l_i^0} + (\cos\alpha_{iz})\frac{(U_6^i - U_3^i)}{l_i^0}\ . \tag{3.4.24}$$

For a small strain, i.e. $|2\varepsilon| \ll 1$,

$$l_i' = l_i^0\sqrt{1+2\varepsilon} = l_i^0(1+\varepsilon)\ . \tag{3.4.25}$$

By considering equation (3.4.24), equation (3.4.19) takes the form

$$\cos\alpha_{ix} = \frac{l_i^0\cos\alpha_{ix} + (U_4^i - U_1^i)}{l_i^0(1+\varepsilon)}\ . \tag{3.4.26}$$

For a small strain,

$$\cos\alpha_{ix} = \frac{[l_i^0\cos\alpha_{ix} + (U_4^i - U_1^i)](1-\varepsilon)}{l_i^0(1+\varepsilon)(1-\varepsilon)}$$

$$= \frac{l_i^0\cos\alpha_{ix} - l_i(\cos\alpha_{ix})\varepsilon + U_4^i - U_1^i}{l_i^0}\ . \tag{3.4.27}$$

By using equation (3.4.24), equation (3.4.27) takes the form

$$\cos\alpha_{ix}' = \cos\alpha_{ix} + (\sin^2\alpha_{ix})\frac{(U_4^i - U_1^i)}{l_i^0}$$

$$-(\cos\alpha_{iy}\cos\alpha_{ix})\frac{(U_5^i - U_2^i)}{l_i^0} - (\cos\alpha_{ix}\cos\alpha_{iz})\frac{(U_6^i - U_3^i)}{l_i^0}\ . \tag{3.4.28}$$

Equation (3.4.28) indicates that $\cos\alpha_{ix}'$ is equal to $\cos\alpha_{ix}$ and a *linear combination* of the nodal displacements. For a small strain the first term in equation (3.4.28) is much larger than the rest of the terms.

By using equation (3.4.28), matrix **G** takes the form

$$\mathbf{G} = \mathbf{A} + \mathbf{O}. \tag{3.4.29}$$

The elements of matrix $\mathbf{O}$ are associated with the last three terms of equation (3.4.28). For a small strain, the magnitude of these terms is much smaller than the magnitude of the elements of matrix $\mathbf{A}$. In most conventional structures the effect of the small elements of matrix $\mathbf{O}$ can be ignored in comparison with the elements of matrix $\mathbf{A}$. Because of the nature of matrix $\mathbf{A}$ in an under-constrained cable net, the elements of matrix $\mathbf{O}$ are important for the analysis of their behaviour.

For the cabled structure shown in Fig. 3.2.1(b), matrix $\mathbf{O}$ takes the form

$$\mathbf{O} = \left[\begin{array}{ccccccc|c}
\multicolumn{7}{c|}{\mathbf{O}_{11}} & \mathbf{O}_{12} \\
\frac{\sqrt{2}}{4a}(U_1+U_2) & 0 & 0 & \frac{U_5-U_1}{c} & 0 & 0 & 0 & 0 \\
\frac{\sqrt{2}}{4a}(U_2-U_1) & & 0 & 0 & 0 & 0 & 0 & 0 \\
0 & \frac{\sqrt{2}}{4a}(U_3-U_4) & \frac{U_3-U_1}{c} & 0 & 0 & 0 & 0 & 0 \\
0 & \frac{U_4-U_2}{b} & \frac{\sqrt{2}}{4a}(U_4-U_3) & 0 & 0 & 0 & 0 & 0 \\
0 & 0 & 0 & \frac{U_5-U_1}{c} & 0 & \frac{\sqrt{2}}{2a}(U_5-U_6) & 0 & 0 \\
0 & 0 & 0 & 0 & 0 & \frac{\sqrt{2}}{4a}(U_6-U_5) & \frac{U_6 \quad U_8}{b} & 0 \\
0 & 0 & 0 & 0 & \frac{U_7-U_8}{c} & 0 & 0 & \frac{\sqrt{2}}{4a}(U_7+U_8) \\
\hline
0 & 0 & 0 & 0 & 0 & 0 & \frac{U_8-U_6}{b} & \frac{\sqrt{2}}{4a}(U_7+U_8) \\
\multicolumn{7}{c|}{\mathbf{O}_{21} \qquad\qquad\qquad \mathbf{O}_{22}} &
\end{array}\right] . \tag{3.4.30}$$

Matrix $\mathbf{O}$ can be partitioned into $\mathbf{O}_{11}$, $\mathbf{O}_{12}$, $\mathbf{O}_{21}$ and $\mathbf{O}_{22}$ in accordance with the partitioning of matrix $\mathbf{A}$. In the case of matrix $\mathbf{O}$ given by equation (3.4.30), $\mathbf{O}_{11}$, $\mathbf{O}_{12}$, $\mathbf{O}_{21}$ and $\mathbf{O}_{22}$ take the form shown there.

The elements of matrix $\mathbf{O}$ which are different from zero for the cable net shown in Fig. 3.2.3 are given in Table 3.4.1 (the displacements are measured in metres).

It is possible to use the product $\mathbf{OP}$ to define a new matrix $\mathbf{V}$. The relationship between matrices $\mathbf{OP}$ and $\mathbf{VU}$ takes the form

$$\mathbf{VU} = \mathbf{OP} \; . \tag{3.4.31}$$

In the case where matrix $\mathbf{O}$ takes the form given by equation (3.4.30) and where the prestressing forces take the magnitude given in Fig. 3.3.3, matrix $\mathbf{V}$ takes the form

Table 3.4.1

$O_{11} = +0.224U_1$	$O_{14} = +0.029U_1 + 0.077U_3$	$O_{15} = 0.004(U_1 - U_4) - 0.03(U_6 - U_3)$	$O_{18} = 0.25(U_1 - U_{10})$
$O_{21} = +0.045U_2 + 0.09U_3$	$O_{24} = +0.234U_2$	$O_{25} = 0.248(U_2 - U_5)$	
$O_{31} = +0.179U_3 + 0.09U_2$	$O_{34} = +0.205U_3 + 0.077U_1$	$O_{35} = 0.244(U_2 - U_5) - 0.03(U_4 - U_1)$	$O_{38} = 0.25(U_3 - U_{12})$
$O_{42} = +0.244U_4$	$O_{45} = +0.004(U_4 - U_1) + 0.03(U_6 - U_3)$	$O_{46} = 0.004(U_4 - U_7) - 0.03(U_6 - U_9)$	$O_{49} = 0.25(U_4 - U_{14})$
$O_{510} = -0.25(U_{16} - U_7)$	$O_{513} = +0.249(U_{17} - U_{14})$	$O_{514} = +0.234U_{10}$	$O_{517} = +0.045U_{10} - 0.09U_{18}$
$O_{62} = +0.179U_6 + 0.09U_5$	$O_{65} = +0.244(U_6 + U_3) + 0.03(U_4 - U_1)$	$O_{66} = 0.244(U_6 - U_9) - 0.03(U_4 - U_7)$	$O_{619} = 0.25(U_6 - U_{15})$
$O_{73} = +0.224U_7$	$O_{76} = +0.004(U_7 - U_4) + 0.03(U_6 - U_9)$	$O_{77} = +0.029U_7 - 0.077U_9$	$O_{710} = 0.25(U_7 - U_{16})$
$O_{83} = +0.045U_8 + 0.098U_9$	$O_{86} = +0.248(U_8 - U_5)$	$O_{87} = 0.234U_8$	
$O_{93} = +0.179U_9 + 0.09U_8$	$O_{96} = +0.244(U_9 - U_6) + 0.03(U_4 - U_7)$	$O_{97} = 0.205U_9 - 0.077U_7$	$O_{910} = 0.25(U_9 - U_{18})$
$O_{108} = +0.25(U_{10} - U_1)$	$O_{1011} = +0.029U_{10} + 0.077U_{12}$	$O_{1012} = +0.004(U_{13} - U_{10}) - 0.03(U_{15} - U_{12})$	$O_{1015} = +0.224\ U_{10}$
$O_{1111} = +0.234U_{11}$	$O_{1112} = +0.248(U_{11} - U_{14})$	$O_{1115} = +0.045U_{11} - 0.09U_{12}$	
$O_{118} = +0.25(U_{12} - U_3)$	$O_{1211} = +0.205U_{12} + 0.077U_{10}$	$O_{1212} = 0.224(U_{12} - U_{15}) - 0.03(U_{13} - U_{10})$	$O_{1215} = +0.179U_{12} - 0.09U_{11}$
$O_{139} = +0.25(U_{13} - U_{14})$	$O_{1312} = +0.004(U_{13} - U_{10}) + 0.03(U_{15} - U_{12})$	$O_{1313} = +0.004(U_{16} - U_{13}) - 0.03(U_{18} - U_{15})$	$O_{1316} = +0.224U_{13}$
$O_{1412} = 0.248(U_{14} - U_{11})$	$O_{1413} = +0.248(U_{14} - U_{17})$	$O_{1416} = +0.045U_{14} - 0.09U_{15}$	
$O_{159} = +0.25(U_{15} - U_6)$	$O_{1512} = +0.244(U_{15} - U_{12}) + 0.03(U_{13} - U_{10})$	$O_{1513} = +0.244(U_{18} - U_{15}) - 0.03(U_{13} - U_{16})$	$O_{1516} = +0.179U_{15} - 0.09U_{14}$
$O_{1610} = +0.25(U_{16} - U_7)$	$O_{1613} = +0.004(U_{16} - U_{13}) + 0.03(U_{15} - U_{18})$	$O_{1614} = +0.029U_{16} - 0.077U_{18}$	$O_{1617} = +0.224U_{16}$
$O_{172} = +0.045U_6 + 0.09U_5$	$O_{175} = +0.249(U_5 - U_2)$	$O_{176} = -0.248(U_8 - U_5)$	
$O_{1810} = +0.25(U_{18} - U_9)$	$O_{1813} = +0.244(U_{18} - U_{15}) + 0.03(U_{13} - U_{16})$	$O_{1814} = +0.205U_{18} - 0.077U_{16}$	$O_{1817} = -0.179U_{18} - 0.09U_{17}$

$$
\mathbf{V} = \left[\begin{array}{ccccccc|c}
\frac{\sqrt{25}}{2}\left(\frac{1}{2a}+\frac{1}{c}\right) & \frac{\sqrt{25}}{4a} & 0 & 0 & -\frac{\sqrt{25}}{2c} & 0 & 0 & 0 \\
\frac{\sqrt{25}}{4a} & \frac{\sqrt{2}}{2}\left(\frac{1}{2a}+\frac{1}{b}\right) & 0 & \frac{\sqrt{52}}{2b} & 0 & 0 & 0 & 0 \\
0 & 0 & \frac{\sqrt{25}}{2}\left(\frac{1}{2a}+\frac{1}{c}\right) & -\frac{\sqrt{25}}{4a} & 0 & 0 & -\frac{\sqrt{25}}{2c} & 0 \\
0 & -\frac{\sqrt{25}}{2b} & -\frac{\sqrt{25}}{4a} & \frac{\sqrt{25}}{2}\left(\frac{1}{a}+\frac{1}{b}\right) & 0 & 0 & 0 & 0 \\
-\frac{\sqrt{25}}{2c} & 0 & 0 & 0 & \frac{\sqrt{25}}{8a} & -\frac{\sqrt{25}}{a}0 & 0 & 0 \\
0 & 0 & 0 & 0 & -\frac{\sqrt{25}}{4c} & \frac{\sqrt{25}}{a}\left(\frac{1}{2a}+\frac{1}{c}\right) & 0 & -\frac{\sqrt{25}}{2b} \\
0 & 0 & \frac{\sqrt{25}}{2c} & 0 & 0 & 0 & \frac{\sqrt{25}}{a}\left(\frac{1}{2a}+\frac{1}{c}\right) & \frac{\sqrt{25}}{4a} \\
\hline
0 & 0 & 0 & 0 & 0 & -\frac{\sqrt{25}}{2c} & \frac{\sqrt{25}}{4a} & \frac{\sqrt{25}}{2}\left(\frac{1}{b}+\frac{1}{2a}\right)
\end{array}\right] .
$$

(upper-left block $\mathbf{V}_{11}$, upper-right $\mathbf{V}_{12}$, lower-left $\mathbf{V}_{21}$, lower-right $\mathbf{V}_{22}$)

(3.4.32)

The elements of matrix **V** have the dimensions of force per length. The dimension of force is that used to determine the prestressing force and the dimension of length is that used to determine the dimensions of the structure.

Also matrix **V** can be partitioned into $\mathbf{V}_{11}$, $\mathbf{V}_{12}$, $\mathbf{V}_{21}$ and $\mathbf{V}_{22}$ as given by equation (3.4.32).

By using equation (3.4.29) and (3.4.31), equation (3.4.18) takes the form

$$\mathbf{AP} + \mathbf{VU} = \mathbf{0} \ . \tag{3.4.33}$$

The contribution of **AP** is much larger than the contribution of **VU** to equation (3.4.33). In every case where the combination of **AP** and **VU** has to be considered in the same equation, the contribution of **VU** can be ignored. The elements of the eighth row of matrix **A** of the net shown in Fig. 3.2.1(b) is a linear combination of the elements of the other rows. In this case it is possible to form an equation with the contribution of matrix **VU** only. This equation takes the form

$$(\mathbf{A}_{21} - \mathbf{W}\mathbf{A}_{11})\mathbf{P}_1 + (\mathbf{A}_{22} - \mathbf{W}\mathbf{A}_{12})\mathbf{P}_2 + (\mathbf{V}_{21} - \mathbf{W}\mathbf{V}_{11})\mathbf{V}_1 + (\mathbf{V}_{22} - \mathbf{W}\mathbf{V}_{12})\mathbf{U}_2 = \mathbf{0} \ . \tag{3.4.34}$$

The fact that the elements of matrix **A** are not independent implies the relationship given by equation (3.2.20). By using the relationship, equation (3.4.34) takes the form

$$(\mathbf{V}_{21} - \mathbf{W}\mathbf{V}_{11})\mathbf{V}_1 + (\mathbf{V}_{12} - \mathbf{W}\mathbf{V}_{12})\mathbf{U}_2 = \mathbf{0} \ . \tag{3.4.35}$$

For the cable net shown in Fig. 3.2.1(b), equation (3.4.35) gives one equation only. This equation can be added to the set of equations given by equation (3.4.9) to form a set of nine equations with nine unknowns. In this way the actual nodal displacements due to the prestressing can be predicted.

For the prestressed tensegric shell shown in Fig. 3.3.4, equation (2.3.8) is a set of 11 equations with 13 unknowns. The unknowns consist of V_1 to V_{12} and Δ_i of the member which is shortened or lengthened in the prestressing. Because of the nature of matrix **A**, it is possible to replace U_1 to U_{12} with new variables e_1 to e_{10} which are presented by vector **e**. Vector **e** can be established by using equation (3.4.13) where **W** is given by equation (3.2.27). In this case, vector $\mathbf{U}_1$ is associated with the equilibrium conditions in the U_1, U_2, U_3, U_4, U_5, U_6, U_7, U_{10}, U_{11} and U_{12} directions. Vector $\mathbf{U}_2$ is associated with the equilibrium conditions in the U_8 and U_9 directions. $\mathbf{U}$, $\mathbf{U}_1$ and $\mathbf{U}_2$ take the forms

$$\mathbf{U} = \begin{bmatrix} \mathbf{U}_1 \\ \mathbf{U}_2 \end{bmatrix} , \quad \mathbf{U}_1 = \begin{bmatrix} U_1 \\ U_2 \\ U_3 \\ U_4 \\ U_5 \\ U_6 \\ U_7 \\ U_{10} \\ U_{11} \\ U_{12} \end{bmatrix} , \quad \mathbf{U}_2 = \begin{bmatrix} U_8 \\ U_9 \end{bmatrix} . \tag{3.4.36}$$

By using vector **e** equation (2.3.8) takes the form of a set of 11 equations with 11 unknowns. By using it, the actual shortening or lengthening of number i associated with the prestressing of the structure can be determined. The nodal displacement caused by the prestressing can be determined by establishing matrix **V** of the structure. By using equation (3.4.35) the additional two equations necessary to complete equation (2.3.8) to a set of 13 equations with 13 unknowns can be established. By using this equation the explicit nodal displacements caused by the prestressing can be determined.

For the cable net shown in Fig. 3.3.5, equation (2.3.8) is a set of 16 equations with 19 unknowns. The unknowns consist of U_1 to U_{18} and Δ_i of the member which is shortened in the prestressing. In this case it is possible to replace U_1 to U_{18} with e_1 to e_{16}. Vector $\mathbf{e}$ can be established by using equation (3.4.13), where $\mathbf{W}$ is given by equation (3.2.30). In this case, $\mathbf{U}_1$ is associated with equilibrium conditions in the directions of U_1, U_2, U_3, U_4, U_{17}, U_6, U_7, U_8, U_9, U_{10}, U_{11}, U_{12}, U_{13}, U_{14}, U_{15} and U_{16}. $\mathbf{U}_2$ is associated with the equilibrium conditions in the directions of U_5 and U_{18}. In this case, $\mathbf{U}$, $\mathbf{U}_1$ and $\mathbf{U}_2$ take the forms

$$\mathbf{U} = \begin{bmatrix} \mathbf{U}_1 \\ \mathbf{U}_2 \end{bmatrix}, \quad \mathbf{U}_1 = \begin{bmatrix} U_1 \\ U_2 \\ U_3 \\ U_4 \\ U_{17} \\ U_6 \\ U_7 \\ U_8 \\ U_9 \\ U_{10} \\ U_{11} \\ U_{12} \\ U_{13} \\ U_{14} \\ U_{15} \\ U_{16} \end{bmatrix}, \quad \mathbf{U}_2 = \begin{bmatrix} U_5 \\ U_{18} \end{bmatrix}. \tag{3.4.37}$$

By using vector $\mathbf{e}$, equation (2.3.8) takes the form of a set of 17 equations with 17 unknowns. By using it the actual shortening associated with prestressing of the cable net can be determined. The explicit 18 nodal displacements due to prestressing can be found by considering the set of 17 equations and establishing matrix $\mathbf{V}$ and using the additional two equations given by equation (3.4.35) to form a set of 19 equations with 19 unknowns.

In some cases, conditions of symmetry can be used to determine the nodal displacements, due to prestressing, explicitly by using equation (2.2.9) only. For example, the net shown in Fig. 3.2.3 is prestressed to the level shown in Fig. 3.3.5 by shortening member 9. Symmetry implies the nodal displacements shown in Fig. 3.4.2.

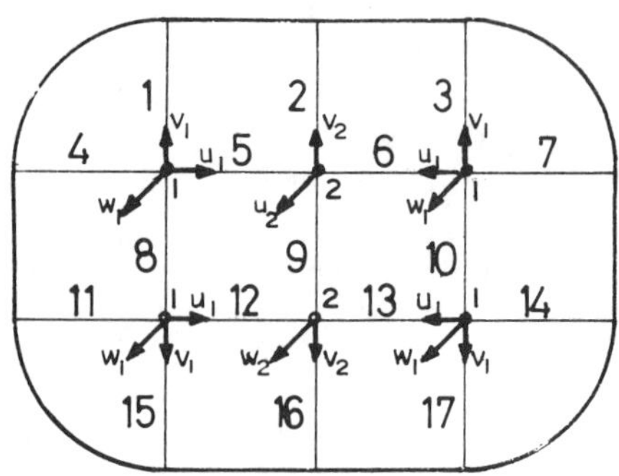

Fig. 3.4.2 — The nodal displacements due to the shortening of element 9.

In the case where all elements are of the same cross-sectional area A_0 and are made of the same material with elasticity modulus E, matrix $\mathbf{A}$ can be used to establish the following equations:

$$P_0 = P_1 = P_3 P_{15} P_{17} = \frac{EA_0}{a}(-0.8v_1 + 0.9w_1)$$

$$P_0 = P_2 = P_{16} = \frac{EA_0}{a}(-0.8v_2 + 0.4w_2)$$

$$1.93P_0 = P_4 = P_7 = P_{11} = P_{14} = \frac{EA_0}{a}(0.9v_1 - 0.5w_1) \tag{3.4.38}$$

$$1.85P_0 = P_5 = P_6 = P_{12} = P_{13} = \frac{EA_0}{a}(-1.0v_1 + 0.1w_1 - 0.1w_2)$$

$$0.91P_0 = P_8 = P_{10}\frac{EA_0}{a}(0.5v_1)$$

in which $P_0 = 20$ kN and $a = 4.0$ m.

Equation (3.4.38) is a consistent set of equations which can be used to determine u_1, v_1, w_1, v_2 and w_2.

$$u_1 = \frac{4.46P_0 a}{EA_0}$$

$$v_1 = \frac{1.82P_0 a}{EA_0}$$

$$w_1 = \frac{6.14P_0 a}{EA_0} \tag{3.4.39}$$

$$v_2 = -\frac{23.21P_0 a}{EA_0}$$

$$w_2 = -\frac{43.9P_0 a}{EA_0}\,.$$

The shortening of element 9 due to prestressing is composed of the change Δ_9 in length due to the nodal displacements given by

$$\Delta_9 = 2v_2 = \frac{46.42P_0a}{EA_0} \tag{3.4.40}$$

and the change Δ_9' in length due to the prestressing force, i.e.

$$\Delta_9' = \frac{0.91P_0a}{EA_0} . \tag{3.4.41}$$

The table shortening of member 9 associated with the prestressing of the cable net is

$$\bar{\Delta}_9 = \frac{47.33P_0a}{EA_0} . \tag{3.4.42}$$

4

The effect of external load

4.1 A FULLY CONSTRAINED CABLED STRUCTURE

The external load, applied to the free nodes of the structure, is presented by vector **Q**. The elements of the vector indicate the load applied to the nodes in the x, y and z directions. Vector **Q** is arranged similarly to vector **U**. Where element U_i of vector **U** indicates a displacement in a given direction, element Q_i of vector **Q** indicates the external load in this direction. The geometrical presentation of the elements of vectors **U** and **Q** of cabled structures are shown in Figs. 3.2.1, 3.2.2 and 3.2.3.

The external load causes a change in the internal forces acting on the members of the structure, and displacements of the free nodes. In the case where the cables of the structure are kept straight, because of prestressing, in all loading cases the equilibrium conditions of statics of the free nodes take the form

$$\mathbf{A}(\mathbf{P}+\mathbf{F})=\mathbf{Q} \tag{4.1.1}$$

where **F** indicates the change in the internal forces acting on the members of the structure where load **Q** is applied to the structure. By using equation (3.2.1), equation (4.1.1) takes the form

$$\mathbf{A}\mathbf{F}=\mathbf{Q}\ . \tag{4.1.2}$$

For a fully constrained cabled structure there are more elements in vector **F** than there are equilibrium equations. In this case it is impossible to determine the change in internal forces due to the external load by using the equilibrium conditions of statics given by equation (4.1.2) only. It is also necessary to consider the nodal displacements. The relationship between the nodal displacements **U** and the change **Δ** in length of the elements takes the form given by equation (2.3.8). The relationship between the change in internal forces and the nodal displacements is given by equation (2.4.3). It can be seen that for a fully constrained cabled structure the analysis follows the method used to analyse common indeterminate reticulated structures. The nodal displacements and the change in internal forces of the members can be analysed by using equation (2.5.5). The only difference between a common reticulated structure and a cabled structure is that in the latter the nodal displacements are measured from the configuration of the prestressed structure.

For example, the analysis of the nodal displacements and the change in internal forces of the members of the tensegric shell shown in Fig. 3.2.2(a) is exactly the same as the analysis of the indeterminate structure shown in Fig. 2.1.2, described in section 2.5.

4.2 AN UNDER-CONSTRAINED CABLED STRUCTURE: THE FITTED-LOAD CASE

For an under-constrained cabled structure the number of equilibrium equations in equation (4.1.2) is equal to or larger than the number of unknown elements of vector **F**. It is possible to partition **F** and **Q** into $\mathbf{F}_1$, $\mathbf{F}_2$ and $\mathbf{Q}_1$, $\mathbf{Q}_2$, respectively, in accordance with $\mathbf{A}_{11}$, $\mathbf{A}_{12}$, $\mathbf{A}_{21}$ and $\mathbf{A}_{22}$. $\mathbf{F}_1$ and $\mathbf{Q}_1$ are the elements of **F** and **Q** associated with $\mathbf{A}_{11}$ and $\mathbf{A}_{12}$. $\mathbf{F}_2$ and $\mathbf{Q}_2$ are associated with $\mathbf{A}_{21}$ and $\mathbf{A}_{22}$. For example, for the cable net shown in Fig. 3.2.1(b) they take the forms

$$\mathbf{F}_1 = \begin{bmatrix} F_1 \\ F_2 \\ F_3 \\ F_4 \\ F_5 \\ F_6 \\ F_7 \end{bmatrix}, \quad \mathbf{F}_2 = [F_8]; \quad \mathbf{Q}_1 = \begin{bmatrix} Q_1 \\ Q_2 \\ Q_3 \\ Q_4 \\ Q_5 \\ Q_6 \\ Q_7 \end{bmatrix}, \quad \mathbf{Q}_2 = [Q_8] \,. \tag{4.2.1}$$

For the tensegric shell shown in Fig. 3.2.2(b) they take the forms

$$\mathbf{F}_1 = \begin{bmatrix} F_1 \\ F_2 \\ F_3 \\ F_4 \\ F_5 \\ F_6 \\ F_7 \\ F_8 \\ F_9 \\ F_{10} \end{bmatrix}, \quad \mathbf{F}_2 = [F_{11}]; \quad \mathbf{Q}_1 = \begin{bmatrix} Q_1 \\ Q_2 \\ Q_3 \\ Q_4 \\ Q_5 \\ Q_6 \\ Q_7 \\ Q_{10} \\ Q_{11} \\ Q_{12} \end{bmatrix}, \quad \mathbf{Q}_2 = \begin{bmatrix} Q_8 \\ Q_9 \end{bmatrix} \,. \tag{4.2.2}$$

For the cable net shown in Fig. 3.2.3 they take the forms

$$\mathbf{F}_1 = \begin{bmatrix} F_1 \\ F_2 \\ F_3 \\ F_4 \\ F_5 \\ F_6 \\ F_7 \\ F_8 \\ F_9 \\ F_{10} \\ F_{11} \\ F_{12} \\ F_{13} \\ F_{14} \\ F_{15} \\ F_{16} \end{bmatrix}, \ \mathbf{F}_2 = [F_{17}]; \ \mathbf{Q}_1 = \begin{bmatrix} Q_1 \\ Q_2 \\ Q_3 \\ Q_4 \\ Q_{17} \\ Q_6 \\ Q_7 \\ Q_8 \\ Q_9 \\ Q_{10} \\ Q_{11} \\ Q_{12} \\ Q_{13} \\ Q_{14} \\ Q_{15} \\ Q_{16} \end{bmatrix}, \ \mathbf{Q}_2 = \begin{bmatrix} Q_5 \\ Q_8 \end{bmatrix}. \tag{4.2.3}$$

By using $\mathbf{F}_1$, $\mathbf{F}_2$, $\mathbf{Q}_1$ and $\mathbf{Q}_2$, equation (4.1.2) takes the form

$$\begin{aligned} \mathbf{A}_{11}\mathbf{F}_1 + \mathbf{A}_{12}\mathbf{F}_2 &= \mathbf{Q}_1 \\ \mathbf{A}_{21}\mathbf{F}_1 + \mathbf{A}_{22}\mathbf{F}_2 &= \mathbf{Q}_2 \, . \end{aligned} \tag{4.2.4}$$

By using equation (3.2.20), equation (4.2.4) takes the form

$$\begin{aligned} \mathbf{A}_{11}\mathbf{F}_1 + \mathbf{A}_{12}\mathbf{F}_2 &= \mathbf{Q}_1 \\ \mathbf{W}(\mathbf{A}_{11}\mathbf{F}_1 + \mathbf{A}_{12}\mathbf{F}_2) &= \mathbf{Q}_2 \, . \end{aligned} \tag{4.2.5}$$

Equation (4.2.5) indicates that the left-hand side of the second equation is a linear combination of the elements on the left-hand side of the first equation. Only in the case where

$$\mathbf{Q}_2 = \mathbf{W}\mathbf{Q}_1 \tag{4.2.6}$$

are the two equations, given in equation (4.2.5), *consistent*; where the first equation of equation (4.2.5) is satisfied, the second is satisfied automatically. The load presented by equation (4.2.6), which indicates a specific relationship between the elements of the vector of the applied load, is called the **fitted load**. For example, for the cable net shown in Fig. 3.2.1(b) the loading case shown in Fig. 4.2.1 is a fitted

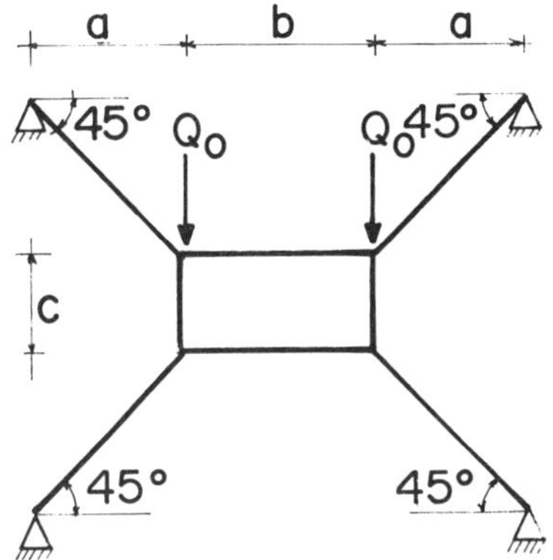

Fig. 4.2.1 — A cable net loaded by a fitted load.

load. It can be seen that condition (4.2.6) is satisfied:

$$0 = [1 \quad 1 \quad 1 \; -1 \; -1 \quad 1 \; -1] \begin{bmatrix} 0 \\ -Q_0 \\ 0 \\ -Q_0 \\ 0 \\ 0 \\ 0 \end{bmatrix}. \tag{4.2.7}$$

For the tensegric shell shown in Fig. 3.2.2(b) the loading case shown in Fig. 4.2.2

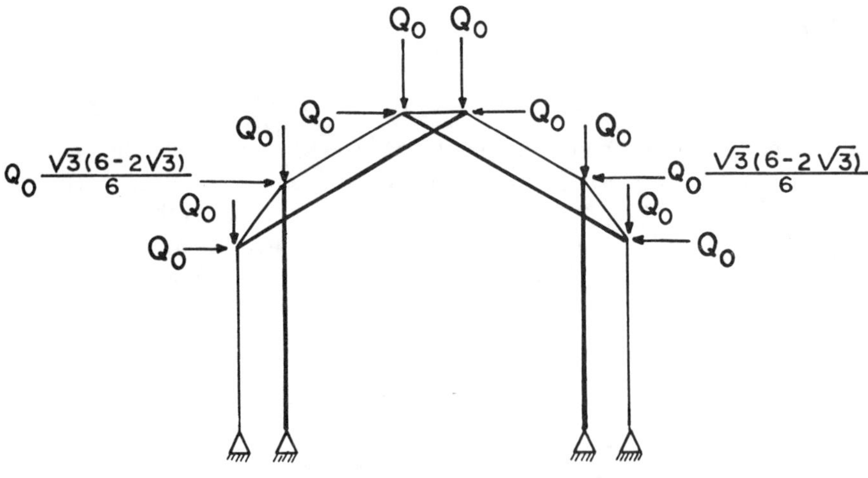

Fig. 4.2.2 — A tensegric shell loaded by a fitted load.

is a fitted load. In this case, equation (4.2.6) applies and the following two equations are satisfied:

$$\begin{bmatrix} -Q_0 \\ -Q_0\dfrac{\sqrt{3}}{6}(6-2\sqrt{3}) \end{bmatrix} = \begin{bmatrix} -\dfrac{2\sqrt{3}}{3} & 0 & -\dfrac{2\sqrt{3}}{3} & 0 & -\dfrac{\sqrt{3}}{3} & -1 & -\dfrac{\sqrt{3}}{3} & 0 & 0 & 0 \\ -1 & 0 & -1 & 0 & -1 & 0 & -1 & 0 & -1 & 0 \end{bmatrix} \begin{bmatrix} Q_0 \\ -Q_0 \\ -Q_0 \\ Q_0 \\ -Q_0 \\ -Q_0 \\ -Q_0 \\ -Q_0 \\ -Q_0 \end{bmatrix} . \tag{4.2.8}$$

For the cable net shown in Fig. 3.2.3 the loading case shown in Fig. 4.2.3 is a fitted load. By using equation (3.2.30), equation (4.2.6) applies and the following two equations are satisfied:

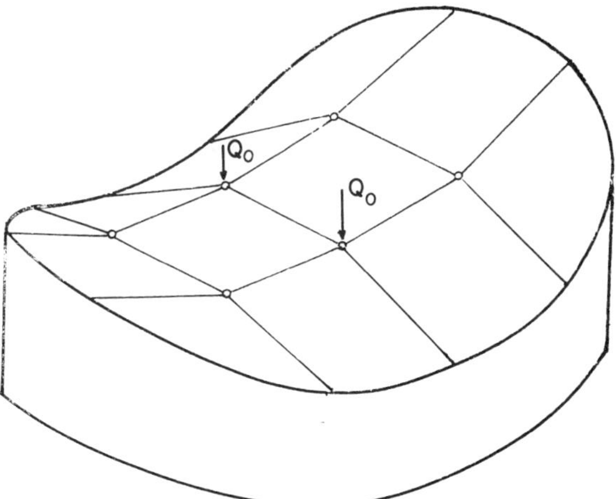

Fig. 4.2.3 — A cable net loaded by a fitted load.

$$\begin{bmatrix} 0 \\ 0 \end{bmatrix} = \mathbf{W} \begin{bmatrix} 0 \\ 0 \\ 0 \\ 0 \\ 0 \\ Q_0 \\ 0 \\ 0 \\ 0 \\ 0 \\ 0 \\ 0 \\ 0 \\ 0 \\ Q_0 \\ 0 \end{bmatrix} . \tag{4.2.9}$$

For the cable net shown in Fig. 3.2.4 the loading case shown in Fig. 4.2.4 where all nodes are loaded by a point load cutting downwards is a fitted load.

In the fitted-load case the change in internal forces can be found by using the first equation of equation (4.2.4) only:

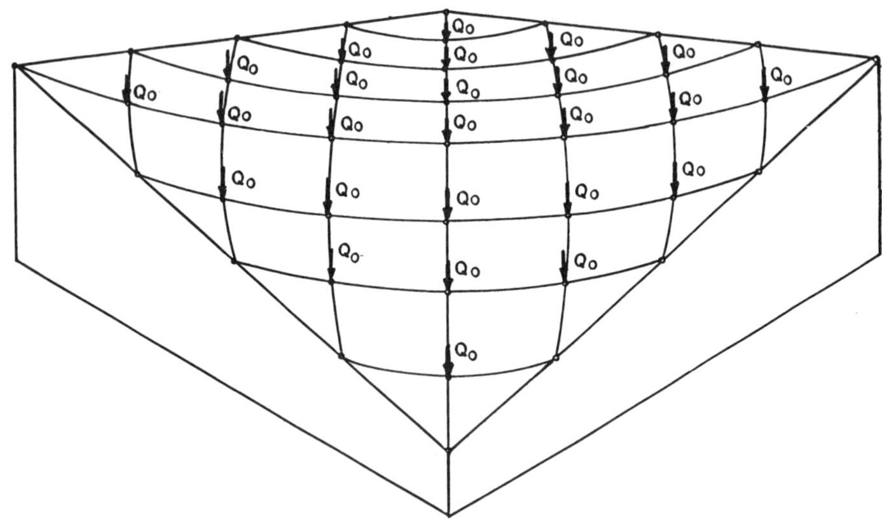

Fig. 4.2.4 — A cable net loaded by a fitted load.

$$\mathbf{A}_{11}\mathbf{F}_1 + \mathbf{A}_{12}\mathbf{F}_2 = \mathbf{Q}_1 \ . \qquad (4.2.10)$$

The second equation of equation (4.2.4) is a linear combination of equation (4.2.10) and does not add any more independent equations to the set. The number of equations in equation (4.2.10) is smaller than the number of unknowns. Thus the change in internal forces of the members of the structure cannot be analysed by considering the equilibrium conditions of statistics at the free nodes only; the nodal displacements should also be considered. The fact that it is possible to determine the change in internal forces by using matrix **A** indicates that the under-constrained cable structure can sustain the fitted load in its prestressed configuration. The nodal displacements in this loading case are due to the elasticity of the members only and they are small. The change in the internal forces due to the change in the length of the members is given by equation (2.4.3). The nodal displacements due to the change in the length of the members takes the form given by equation (2.3.8). By partitioning $\mathbf{U}$ into $\mathbf{U}_1$ and $\mathbf{U}_2$ in accordance with $\mathbf{Q}_1$ and $\mathbf{Q}_2$ and by partitioning $\mathbf{\Delta}$ into $\mathbf{\Delta}_1$ and $\mathbf{\Delta}_2$ in accordance with $\mathbf{F}_1$ and $\mathbf{F}_2$, equation (2.3.8) takes the form given by equation (3.4.10) which by using equation (3.2.20) takes the form given by equation (3.4.12).

By using $\mathbf{e}$ defined by equation (3.4.13), equation (3.4.12) takes the form

$$\begin{aligned} \mathbf{\Delta}_1 &= \mathbf{A}_{11}^{\mathrm{T}}\mathbf{e} \\ \mathbf{\Delta}_2 &= \mathbf{A}_{12}^{\mathrm{T}}\mathbf{e} \ . \end{aligned} \qquad (4.2.11)$$

By using equation (2.4.3), $\mathbf{F}_1$ and $\mathbf{F}_2$ take the forms

$$\begin{aligned}\mathbf{F}_1 &= \mathbf{S}_{11}\mathbf{A}_{11}^{\mathrm{T}}\mathbf{e}\\ \mathbf{F}_2 &= \mathbf{S}_{22}\mathbf{A}_{11}^{\mathrm{T}}\mathbf{e}\ .\end{aligned} \tag{4.2.12}$$

$\mathbf{S}_{11}$ and $\mathbf{S}_{22}$ are the matrices formed from matrix $\mathbf{S}$ in accordance with $\mathbf{A}_{11}$ and $\mathbf{A}_{22}$. The elements on the diagonal of matrix $\mathbf{S}_{11}$ are functions of the properties of the members of the structure associated with $\mathbf{F}_1$. The elements on the diagonal of matrix $\mathbf{S}_{22}$ are functions of the members associated with $\mathbf{F}_2$. By using equation (4.2.12) and equation (4.2.10), $\mathbf{e}$ takes the form

$$\mathbf{e} = (\mathbf{A}_{11}\mathbf{S}_{11}\mathbf{A}_{11}^{\mathrm{T}} + \mathbf{A}_{12}\mathbf{S}_{22}\mathbf{A}_{12}^{\mathrm{T}})^{-1}\mathbf{Q}_1\ . \tag{4.2.13}$$

By using equation (4.2.13), equation (4.2.12) takes the form

$$\begin{aligned}\mathbf{F}_1 &= \mathbf{S}_{11}\mathbf{A}_{11}^{\mathrm{T}}(\mathbf{A}_{11}\mathbf{S}_{11}\mathbf{A}_{11}^{\mathrm{T}} + \mathbf{A}_{12}\mathbf{S}_{22}\mathbf{A}_{12}^{\mathrm{T}})^{-1}\mathbf{Q}_1\\ \mathbf{F}_2 &= \mathbf{S}_{22}\mathbf{A}_{12}^{\mathrm{T}}(\mathbf{A}_{11}\mathbf{S}_{11}\mathbf{A}_{11}^{\mathrm{T}} + \mathbf{A}_{12}\mathbf{S}_{22}\mathbf{A}_{12}^{\mathrm{T}})^{-1}\mathbf{Q}_1\ .\end{aligned} \tag{4.2.14}$$

By using equation (4.2.14), the change in internal forces can be determined.

For the cable net shown in Fig. 4.2.1, $c=b=a$, all cables are made of the same material, the areas of their cross-sections are the same, and the change in the internal forces found by using equation (4.2.14) takes the form shown in Fig. 4.2.5.

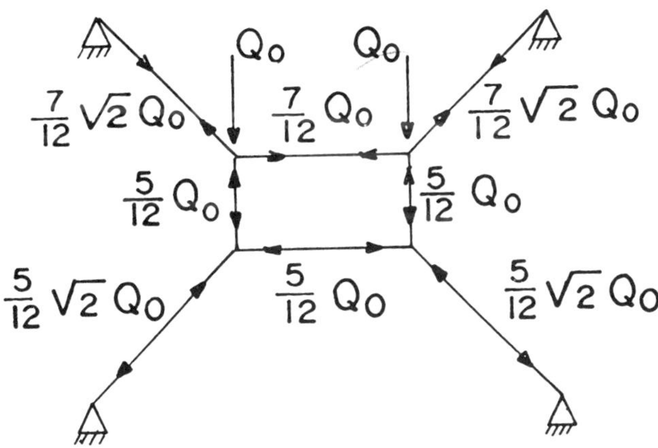

Fig. 4.2.5 — The change in internal forces of the cable net due to a fitted load.

For the cable net loaded by the fitted load shown in Fig. 4.2.3, where all elements have the same cross-sectional area, the change in the internal forces found by using equation (4.2.14) takes the form shown in Fig. 4.2.6.

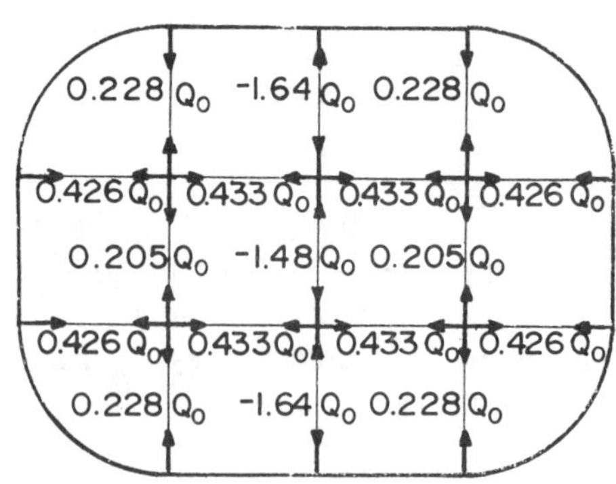

Fig. 4.2.6 — The change in the internal forces due to the fitted load shown in Fig. 4.2.3.

In some cases, simple consideration can be used to determine the internal forces in the structure. For example in the case where the cable net shown in Fig. 3.2.4 is loaded by the fitted load shown in Fig. 4.2.4 and all elements have the same cross-sectional area, it can be assumed that half of the load is carried by the cables in the x direction and half is loaded on the cables in the y direction which are acting as arches in this loading case. The change in the internal forces found by using equation (4.2.14) takes the form shown in Fig. 4.2.7.

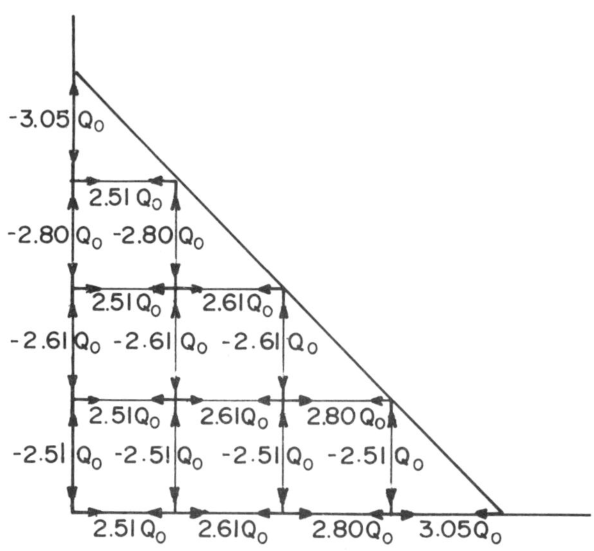

Fig. 4.2.7 — The change in the internal forces due to the fitted load shown in Fig. 4.2.4.

Where the change in internal forces acting on the members of the structure is known, the elements of vector $\mathbf{e}$ can be defined by using the first equation of equation (4.2.12) which takes the form

$$\mathbf{e} = (\mathbf{S}_{11}\mathbf{A}_{11}^{\mathrm{T}})^{-1}\mathbf{F}_1 \; . \tag{4.2.15}$$

Equations (4.2.15) and (3.4.12) are not enough to define the nodal displacements uniquely. In this case the nodal displacements can be determined by considering the geometrical second-order effect. By considering equation (3.4.29), equation (4.1.1) takes the form

$$(\mathbf{A}+\mathbf{O})(\mathbf{P}+\mathbf{F}) = \mathbf{Q} \; . \tag{4.2.16}$$

By using equations (3.2.1), (2.4.3) and (2.3.8), equation (4.2.16) takes the form

$$(\mathbf{ASA}^{\mathrm{T}})\mathbf{U} + \mathbf{OP} + \mathbf{OF} = \mathbf{Q} \; . \tag{4.2.17}$$

In equation (4.2.17) the first and the second terms are a linear combination of the nodal displacements. Where the structure is loaded by a ‘fitted load’, the change in internal forces acting on the members, presented by vector $\mathbf{F}$, is known and is given by equation (4.2.14). In this case the third term of equation (4.2.17) is a linear combination of the nodal displacements. By using equation (4.2.14) it is possible to define matrix $\mathbf{Y}$:

$$\mathbf{OP} + \mathbf{OF} = \mathbf{YU} \; . \tag{4.2.18}$$

By using equation (4.2.18), equation (4.2.17) takes the form

$$(\mathbf{ASA}^{\mathrm{T}})\mathbf{U} + \mathbf{YU} = \mathbf{Q} \; . \tag{4.2.19}$$

Also matrix $\mathbf{Y}$ can be partitioned into $\mathbf{Y}_{11}$, $\mathbf{Y}_{12}$, $\mathbf{Y}_{21}$ and $\mathbf{Y}_{22}$ in accordance with $\mathbf{Q}_1$, $\mathbf{Q}_2$, $\mathbf{U}_1$ and $\mathbf{U}_2$.

In practical cases of cabled structures the contribution of the first term of equation (4.2.19) is much larger than the contribution of the second term. In cases where two terms appear in the equation, the contribution of the second term can be ignored. Because not all the elements of matrix $\mathbf{A}$ are independent, it is possible to form the combination in which the elements of the first term vanish and only the elements of the second terms appear, in which case the equation takes the form

$$(\mathbf{Y}_{21} - \mathbf{WY}_{11})\mathbf{U}_1 + (\mathbf{Y}_{22} - \mathbf{WY}_{12})\mathbf{U}_2 = \mathbf{Q}_2 - \mathbf{WQ}_1 \; . \tag{4.2.20}$$

The equation can be added to equation (4.2.13) to form the set of equations required to determine the nodal displacements

$$\begin{aligned} &\mathbf{U}_1+\mathbf{W}^{\mathrm{T}}\mathbf{U}_2=(\mathbf{A}_{11}\mathbf{S}_{11}\mathbf{A}_{11}^{\mathrm{T}}+\mathbf{A}_{12}\mathbf{S}_{22}\mathbf{A}_{12}^{\mathrm{T}})^{-1}\mathbf{Q}_1 \\ &(\mathbf{Y}_{21}-\mathbf{W}\mathbf{Y}_{11})\mathbf{U}_1+(\mathbf{Y}_{22}-\mathbf{W}\mathbf{Y}_{12})\mathbf{U}_2=0\;. \end{aligned} \tag{4.2.21}$$

The solution of equation (4.2.21) takes the form

$$\begin{aligned} &\mathbf{U}_1=[1-\mathbf{W}\mathbf{E}^{\mathrm{T}}(\mathbf{Y}_{22}-\mathbf{W}\mathbf{Y}^{12})^{-1}(\mathbf{Y}_{21}-\mathbf{W}\mathbf{Y}_{11})]^{-1}(\mathbf{A}_{11}\mathbf{S}_{11}\mathbf{A}_{11}^{\mathrm{T}}+\mathbf{A}_{12}\mathbf{S}_{22}\mathbf{A}_{12}^{\mathrm{T}})^{-1}\mathbf{Q}_1 \\ &\mathbf{U}_2=-(\mathbf{Y}_{22}-\mathbf{W}\mathbf{Y}_{12})^{-1}(\mathbf{Y}_{21}-\mathbf{W}\mathbf{Y}_{11})\mathbf{U}_1\;. \end{aligned} \tag{4.2.22}$$

By using equation (4.2.22), the nodal displacements of a cabled structure loaded with a fitted load can be determined.

For the cable net shown in Fig. 4.2.1, equation (4.2.22) is composed of seven equations of the type of the first equation of equation (4.2.22) and one equation of the second type. The vector of the nodal displacements takes the form given by equation (3.4.14).

For the fitted load shown in Fig. 4.2.1, where $c=b=a$ and all cables have the same elastic modulus E and cross-sectional area A_0, the nodal displacements found by using equation (4.2.22) take the form shown in Fig. 4.2.8.

For the tensegric shell shown in Fig. 4.2.2, equation (4.2.22) is composed of ten equations of the type of the first equation of equation (4.2.22) and two of the type of the second equation of equation (4.2.22). The vector of the nodal displacements takes the form given by equation (3.4.36).

In some cases, symmetry of the structure and of the expected nodal displacements can be used to ease the analysis. It is most helpful where, by using symmetry, an under-constrained cabled structure can be analysed by the simple methods of analysis of a fully constrained cabled structure.

For example, for the cable net loaded by the fitted load shown in Fig. 4.2.3 the cable net is under-constrained; there are 17 elements and 18 nodal displacements and equilibrium equations. The nodal displacements associated with the fitted load can be found by using matrix **Y** and equation (4.2.22). Solving the problem by following this approach requires the cumbersome analysis of matrix **O**. In this case, symmetry of the expected nodal displacement can be of considerable help. Because of the symmetry of the load and the structure the nodal displacements are expected to take the symmetrical pattern shown in Fig. 3.4.2. In this case there are six different members and only five different nodal displacements and equilibrium conditions to satisfy. The set of equilibrium equations is similar to that used to analyse a fully constrained cabled structure. This set of equations is simple to analyse. The nodal displacements associated with the fitted load shown in Fig. 4.2.3, where all elements

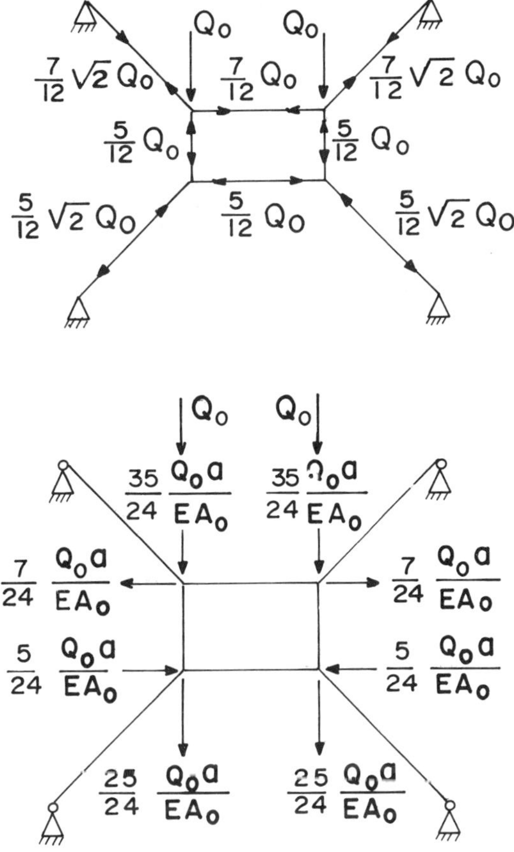

Fig. 4.2.8 — The nodal displacements of the cable net due to a fitted load.

have the same cross-sectional area A_0 and the same elasticity modulus E and where $a = 4.0$ m, are shown in Fig. 4.2.9.

4.3 AN UNDER-CONSTRAINED CABLED STRUCTURE: THE NON-FITTED-LOAD CASE

4.3.1 The equation for a cabled structure

In the case where the load does not satisfy equation (4.2.6), equation (4.2.4) is not a consistent set of linear equations. There is a contradiction between the first equation and the second equation of equation (4.2.4). If $\mathbf{F}_1$ and $\mathbf{F}_2$ satisfy the first equation, it is impossible to satisfy the second equation. The physical implication of this fact is that the prestressed cabled structure cannot sustain the non-fitted load in its prestressed configuration. There will be nodal displacements due to rigid-body movement of the members of the structure. These displacements change the

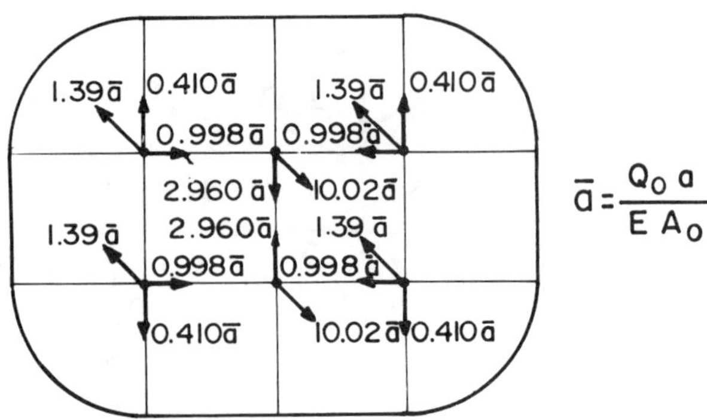

Fig. 4.2.9 — The nodal displacements due to the fitted load shown in Fig. 4.2.3.

configuration of the structure until equilibrium is established at all nodes. The equilibrium of statics at the nodes of the deformed structure takes the form

$$\mathbf{G}(\mathbf{P}+\mathbf{F})=\mathbf{Q}\ . \tag{4.3.1}$$

The elements of the equilibrium matrix **G** are determined by considering the nodal displacements. It has already been shown that matrix **G** can be seen as composed of matrix **A** and matrix **O** as given by equation (3.4.29). The elements of matrix **O** are a linear combination of the nodal displacements. By using equation (3.4.29), equation (4.3.1) takes the form

$$\mathbf{AF}+\mathbf{OP}+\mathbf{OF}=\mathbf{Q}\ . \tag{4.3.2}$$

By using equation (3.4.31), equation (4.3.2) takes the form

$$\mathbf{AF}+\mathbf{VU}+\mathbf{OF}=\mathbf{Q}\ . \tag{4.3.3}$$

There are two types of unknown in equation (4.3.3): internal forces represented by vector **F** and nodal displacements represented by vector **U**. In order to analyse **U** and **F**, it is necessary to establish the relationship between them. In spite of the large nodal displacements, the actual change in the length of the members is small. **F** is related to the change in the length of the bars as given by equation (2.4.3). The relationship between $\mathbf{\Delta}$ and **U** for large nodal displacements is investigated by studying the change in the configuration and length of a typical member of the structure. In the case of member i shown in Fig. 2.2.2, the relationship is given by equation (2.2.7). By using equation (2.2.6) in the case where the second-order effect of the geometrical distortion is not ignored, equation (2.2.3) takes the form

$$\begin{aligned}\Delta_i = {} & (\cos\alpha_{ix})(U_3 + U_1) + (\cos\alpha_{iy})(U_4 - U_1) \\ & + \frac{(U_3 - U_1)^2}{2l_i^0} + \frac{(U_4 - U_2)^2}{2l_i^0} - \frac{(\cos^2\alpha_{ix})(U_3 - U_1)^2}{2l_i^0} \\ & - \frac{(\cos\alpha_{ix}\cos\alpha_{iy})(U_3 - U_1)(U_4 - U_2)}{l_i^0} \\ & - \frac{(\cos^2\alpha_{iy})(U_4 - U_2)^2}{2l_i^0}\,. \end{aligned} \tag{4.3.4}$$

Grouping of equation (4.3.4) takes the form

$$\begin{aligned}\Delta_i = {} & (\cos\alpha_{ix})(U_3 - U_1) + (\cos\alpha_{iy})(U_4 - U_2) \\ & + \frac{(\sin^2\alpha_{ix})(U_3 - U_1)^2}{2l_i^0} + \frac{(\sin^2\alpha_{iy})(U_4 - U_2)}{2l_i^0} \\ & - \frac{(\cos\alpha_{ix}\cos\alpha_{iy})(U_3 - U_1)(U_4 - U_2)}{l_i^0}\,. \end{aligned} \tag{4.3.5}$$

Matrices **A** and **O** of the member of the structure shown in Fig. 2.2.2, considering the load shown in Fig. 4.3.1, take the form

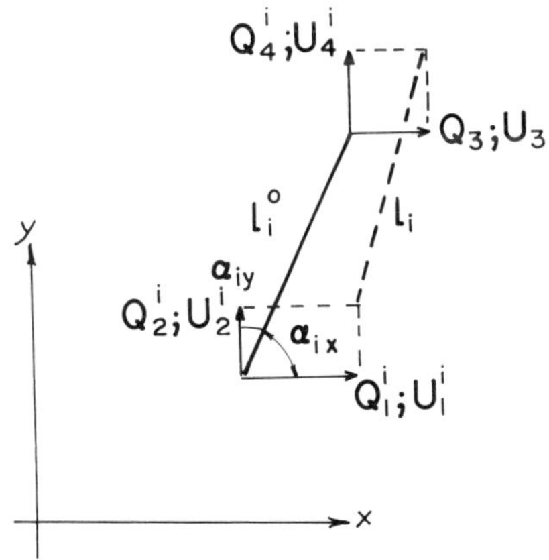

Fig. 4.3.1 — The load of bar *i*.

$$\mathbf{A} = \begin{bmatrix} -\cos\alpha_{ix} \\ -\cos\alpha_{iy} \\ \cos\alpha_{ix} \\ \cos\alpha_{iy} \end{bmatrix} \tag{4.3.6}$$

$$\mathbf{O} = \begin{bmatrix} \dfrac{(\cos\alpha_{iy}\cos\alpha_{ix})(U_4 - U_2)}{l_i^0} + \dfrac{(\sin^2\alpha_{ix})(U_3 - U_1)}{l_i^0} \\ \dfrac{(\cos\alpha_{iy}\cos\alpha_{ix})(U_3 - U_1)}{l_i^0} + \dfrac{(\sin^2\alpha_{iy})(U_4 - U_2)}{l_i^0} \\ \dfrac{(\sin^2\alpha_{ix})(U_3 - U_1)}{l_i^0} + \dfrac{(\cos\alpha_{iy}\cos\alpha_{ix})(U_4 - U_2)}{l_i^0} \\ \dfrac{(\sin^2\alpha_{iy})(U_4 - U_2)}{l_i^0} + \dfrac{(\cos\alpha_{iy}\cos\alpha_{ix})(U_3 - U_1)}{l_i^0} \end{bmatrix} . \tag{4.3.7}$$

The vector of the nodal displacements takes the form

$$\mathbf{U} = \begin{bmatrix} U_1 \\ U_2 \\ U_3 \\ U_4 \end{bmatrix} . \tag{4.3.8}$$

It can be seen that the first two terms on the right-hand side of equation (4.3.5) can be expressed by using matrix **A** and vector **U**:

$$(\cos\alpha_{ix})(U_3 - U_1) + (\cos\alpha_{iy})(U_4 - U_2) = \mathbf{A}^{\mathrm{T}}\mathbf{U} . \tag{4.3.9}$$

This is a method for establishing the relationship between the nodal displacements and the change in the length of the elements of the structure, without using virtual work. The method of virtual work was used in section 2.3.

The last three terms on the right-hand side of equation (4.3.5) can be expressed by using matrices **O** and **U**:

$$\frac{(\sin^2\alpha_{ix})(U_3 - U_1)^2}{2l_i^0} + \frac{(sin^2\alpha_{iy})(U_4 - U_2)^2}{2l_i^0} - \frac{(\cos^2\alpha_{ix}\cos\alpha_{iy})(U_3 - U_1)(U_4 - U_2)}{l_i^0} = \frac{\mathbf{O}^{\mathrm{T}}\mathbf{U}}{2} . \tag{4.3.10}$$

By using equations (4.3.9) and (4.3.10), equation (4.3.5) takes the form

$$\mathbf{\Delta} = (\mathbf{A}^{\mathrm{T}} + \tfrac{1}{2}\mathbf{O}^{\mathrm{T}})\mathbf{U} \; . \tag{4.3.11}$$

Equation (4.3.11) is the generalized form of equation (2.3.8) in which the second-order effect of the nodal displacements is considered. By using equation (4.3.11), equation (2.4.3) takes the form

$$\mathbf{F} = \mathbf{S}(\mathbf{A}^{\mathrm{T}} + \tfrac{1}{2}\mathbf{O}^{\mathrm{T}})\mathbf{U} \; . \tag{4.3.12}$$

By using equation (4.3.12), equation (4.3.3) takes the form

$$\mathbf{DU} + \mathbf{VU} + \tfrac{1}{2}\mathbf{ASO}^{\mathrm{T}}\mathbf{U} + \mathbf{OSA}^{\mathrm{T}}\mathbf{U} + \tfrac{1}{2}\mathbf{OSO}^{\mathrm{T}}\mathbf{U} = \mathbf{Q} \; . \tag{4.3.13}$$

Equation (4.3.13) is the governing equation of cabled structures. It indicates the relationship between the nodal displacements and the load applied to a cabled structure. The first and second terms of equation (4.3.13) are a linear combination of the nodal displacements. The second and third terms are second-order functions of the nodal displacements. In most practical cases the contribution of the first term of equation (4.3.13) is much larger than the contribution of the rest of the terms. The contribution of the second, third, fourth and fifth terms is of the same order of magnitude.

4.3.2 The analysis of under-constrained cabled structures

Equation (4.3.13) is a set of non-linear equations of the nodal displacement. Analysis of a set of non-linear equations is cumbersome. It is important to identify where the linear approximation can be used to reduce the difficulty of analysis as much as possible. In order to explore these possibilities, equation (4.3.11) is investigated. The first term of this equation is a linear combination of the nodal displacements. The second term indicates the second-order effect of the nodal displacements on the change of length of the bars. Where second order is ignored, equation (4.3.11) takes the form

$$\mathbf{\Delta} = \mathbf{A}^{\mathrm{T}}\mathbf{U} \; . \tag{4.3.14}$$

By using $\mathbf{\Delta}_1$, $\mathbf{\Delta}_2$, $\mathbf{A}_{11}$, $\mathbf{A}_{12}$, $\mathbf{A}_{21}$, $\mathbf{A}_{22}$, $\mathbf{U}_1$ and $\mathbf{U}_2$, equation (4.3.14) takes the form

$$\begin{aligned} \mathbf{\Delta}_1 &= \mathbf{A}_{11}^{\mathrm{T}}\mathbf{U}_1 + \mathbf{A}_{21}^{\mathrm{T}}\mathbf{U}_2 \\ \mathbf{\Delta}_2 &= \mathbf{A}_{12}^{\mathrm{T}}\mathbf{U}_1 + \mathbf{A}_{22}^{\mathrm{T}}\mathbf{U}_2 \; . \end{aligned} \tag{4.3.15}$$

By using equation (3.2.20), equation (4.3.15) takes the form

$$\begin{aligned}\mathbf{\Delta}_1 &= \mathbf{A}_{11}^{\mathrm{T}}(\mathbf{U}_1 + \mathbf{W}^{\mathrm{T}}\mathbf{U}_2)\\ \mathbf{\Delta}_2 &= \mathbf{A}_{12}^{\mathrm{T}}(\mathbf{U}_1 + \mathbf{W}^{\mathrm{T}}\mathbf{U}_2)\ .\end{aligned} \tag{4.3.16}$$

It can be seen that in the case where

$$\mathbf{U}_1 = -\mathbf{W}^{\mathrm{T}}\mathbf{U}_2$$

then

$$\mathbf{\Delta}_1 = \mathbf{\Delta}_2 = 0\ . \tag{4.3.17}$$

Equation (4.3.17) indicates that where the nodal displacements take this form the change in the length of members of the structure and in their internal forces is due to the second-order effect of the nodal displacements only. This demonstrates that, by considering the first-order effect of the nodal displacements where they take the form given by equation (4.3.17), the change in the configuration of the structure is due to rigid-body movement of the members of the structure.

It can be seen that displacements given by vector $\mathbf{U}_2$ can be assumed and, by using equation (4.3.17), the nodal displacements given by vector $\mathbf{U}_1$ in the case of rigid-body movement of the members of the structure can be predicted. The number of elements of vector $\mathbf{U}_2$ indicates the degree of freedom of the nodal displacements caused by the rigid-body movement of the members of the structure. For example, for the cable net shown in Fig. 3.2.1(b), the nodal displacements, given by equation (4.3.20) and shown in Fig. 4.3.2, satisfy equation (4.3.17) and are due to rigid-body movement of the members of the structure (it can be seen that in this cable net there is one degree of freedom only):

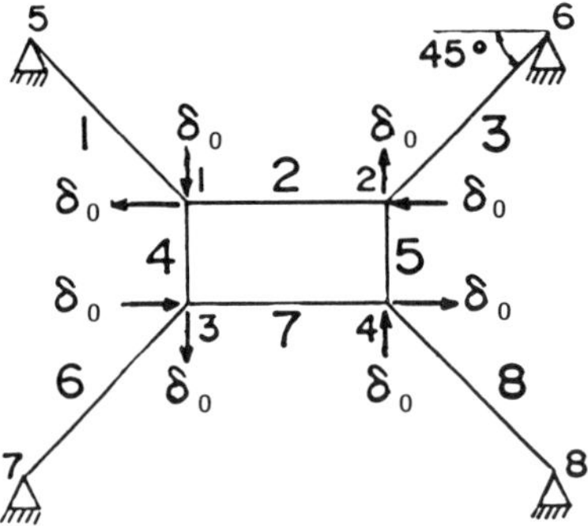

Fig. 4.3.2 — The nodal displacements associated with rigid-body movement of the members of the net.

$$\mathbf{U} = \begin{bmatrix} \mathbf{U}_1 \\ \mathbf{U}_2 \end{bmatrix}, \quad \mathbf{U}_1 = \begin{bmatrix} U_1 \\ U_2 \\ U_3 \\ U_4 \\ U_5 \\ U_6 \\ U_7 \end{bmatrix} = \begin{bmatrix} -1 \\ -1 \\ -1 \\ 1 \\ 1 \\ -1 \\ 1 \end{bmatrix} \delta_0, \quad \mathbf{U}_2 = [U_8] = [\delta_0] \,. \tag{4.3.18}$$

For the tensegric shell shown in Fig. 3.2.2(b) the displacements given by equation (4.3.20) satisfy equation (4.3.17) and are due to rigid-body movement of the members of the structure:

$$\mathbf{U} = \begin{bmatrix} \mathbf{U}_1 \\ \mathbf{U}_2 \end{bmatrix}$$

$$\mathbf{U}_2 = \begin{bmatrix} U_8 \\ U_9 \end{bmatrix} = \begin{bmatrix} \delta_0 \\ \delta_1 \end{bmatrix} \tag{4.3.19}$$

$$\mathbf{U}_1 = \begin{bmatrix} U_1 \\ U_2 \\ U_3 \\ U_4 \\ U_5 \\ U_6 \\ U_7 \\ U_{10} \\ U_{11} \\ U_{12} \end{bmatrix} = \begin{bmatrix} \frac{2\sqrt{3}}{3}\delta_0 + \delta_1 \\ 0 \\ \frac{2\sqrt{3}}{3}\delta_0 + \delta_1 \\ 0 \\ \frac{\sqrt{3}}{3}\delta_0 + \delta_1 \\ \delta_0 \\ \frac{\sqrt{3}}{3}\delta_0 + \delta_1 \\ 0 \\ \delta_1 \\ 0 \end{bmatrix} .$$

It can be seen that in this structure there are two degrees of freedom associated with the rigid-body movement of the members of the structure.

For the cable net shown in Fig. 3.2.3 the displacements given by equation (4.3.21) found by using equations (4.3.17) and (3.2.30) satisfy equation (4.3.17) and are due to rigid-body movement of the members of the structure:

$$\mathbf{U} = \begin{bmatrix} \mathbf{U}_1 \\ \mathbf{U}_2 \end{bmatrix}$$

$$\mathbf{U}_2 = \begin{bmatrix} U_5 \\ U_{18} \end{bmatrix} = \begin{bmatrix} \delta_0 \\ \delta_1 \end{bmatrix}$$

$$\mathbf{U}_1 = \begin{bmatrix} U_1 \\ U_2 \\ U_3 \\ U_4 \\ U_{17} \\ U_6 \\ U_7 \\ U_8 \\ U_9 \\ U_{10} \\ U_{11} \\ U_{12} \\ U_{13} \\ U_{14} \\ U_{15} \\ U_{16} \end{bmatrix} = \begin{bmatrix} -0.749\,\delta_0 + 0.375\,\delta_1 \\ -1.000\,\delta_0 + 0.500\,\delta_1 \\ -2.000\,\delta_0 + 1.000\,\delta_1 \\ -0.249\,\delta_0 + 0.249\,\delta_1 \\ -0.500\,\delta_1 \\ 2.000\,\delta_0 \\ 0.375\,\delta_1 \\ -0.500\,\delta_1 \\ -1.000\,\delta_1 \\ 0.749\,\delta_0 + 0.375\,\delta_1 \\ -1.000\,\delta_0 + 0.500\,\delta_1 \\ 2.000\,\delta_0 - 1.000\,\delta_1 \\ 0.249\,\delta_0 - 0.249\,\delta_1 \\ 1.000\,\delta_0 \\ -2.000\delta_0 \\ -0.375\delta_1 \end{bmatrix} . \tag{4.3.20}$$

Also in this case there are two degrees of freedom associated with the rigid-body movement of the members. The case where $\delta_1 = 0$ is shown in Fig. 4.3.3 and the case where $\delta_0 = 0$ is shown in Fig. 4.3.4.

In the general case, the nodal displacements are assumed to be composed of displacements due to rigid-body movements denoted by $\mathbf{U}_l^0$ and displacements associated with the elastic deformation of the members of the structure denoted by $\mathbf{U}_s^0$:

$$\mathbf{U} = \mathbf{U}_l^0 + \mathbf{U}_s^o \, . \tag{4.3.21}$$

By equation (4.3.17), $\mathbf{U}_l^0$ takes the form

$$\mathbf{U}_l^0 = \begin{bmatrix} -\mathbf{W}^T & \mathbf{U}_l \\ & \mathbf{U}_l \end{bmatrix} \tag{4.3.22}$$

in which $\mathbf{U}_l$ is the vector of the independent nodal displacements necessary to

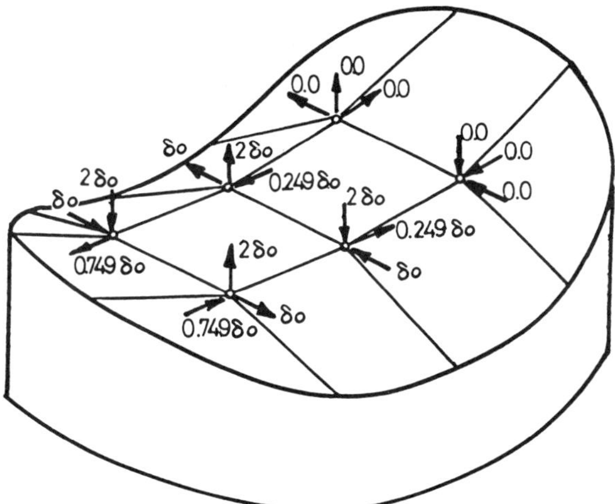

Fig. 4.3.3 — One possible set of nodal displacements due to rigid-body movement of the elements of the cable net.

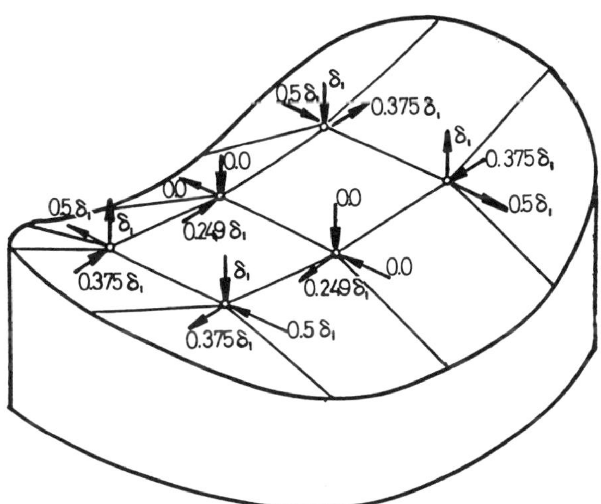

Fig. 4.3.4 — Second possible set of nodal displacements due to rigid-body movement of the elements of the cable net.

determine the nodal displacement due to rigid-body movement of the members of the structure. The order of vector $\mathbf{U}_l$ is equal to the number of rows in matrix $\mathbf{A}_{21}$ or $\mathbf{A}_{22}$ which is equal to $m - r$. $\mathbf{U}_s^0$ takes the form

$$\mathbf{U}_s^0 = \begin{bmatrix} \mathbf{U}_s \\ \boldsymbol{\phi} \end{bmatrix} . \tag{4.3.23}$$

$\mathbf{U}_s$ is a vector of the order equal to the number of rows in matrix $\mathbf{A}_{11}$ or $\mathbf{A}_{12}$ equal to r, and $\boldsymbol{\phi}$ is a zero vector of the order of $\mathbf{U}_l$.

The nodal displacements associated with rigid-body movement are much larger than those associated with the elasticity of the elements:

$$|-\mathbf{W}^T\mathbf{U}_l| \gg |\mathbf{U}_s| . \tag{4.3.24}$$

Equation (4.3.24) implies that, in the analysis of the elements of matrix $\mathbf{V}$ and matrix $\mathbf{O}$, the effect of $\mathbf{U}_s$ can be ignored. By considering $\mathbf{U}_s^0$ and $\mathbf{U}_l^0$, equation (4.3.13) takes the form

$$\mathbf{D}\mathbf{U}_s^0 + \mathbf{V}\mathbf{U}_l^0 + \tfrac{1}{2}\mathbf{A}\mathbf{S}\mathbf{O}^T\mathbf{U}_l^0 + \mathbf{O}\mathbf{S}\mathbf{A}^T\mathbf{U}_s^0 + \tfrac{1}{2}\mathbf{O}\mathbf{S}\mathbf{O}^T\mathbf{U}_l^0 = \mathbf{Q} . \tag{4.3.25}$$

Grouping the elements of equation (4.3.25) takes the form

$$\mathbf{G}\mathbf{U}_s^0 + \mathbf{E}\mathbf{U}_l^0 = \mathbf{Q} \tag{4.3.26}$$

where

$$\begin{aligned} \mathbf{E} &= \mathbf{V} + \tfrac{1}{2}\mathbf{A}\mathbf{S}\mathbf{O}^T + \tfrac{1}{2}\mathbf{O}\mathbf{S}\mathbf{O}^T \\ \mathbf{G} &= \mathbf{D} + \mathbf{O}\mathbf{S}\mathbf{A}^T . \end{aligned} \tag{4.3.27}$$

It is possible to partition $\mathbf{G}$ and $\mathbf{E}$ into $\mathbf{G}_{11}$, $\mathbf{G}_{12}$, $\mathbf{G}_{21}$, $\mathbf{G}_{22}$, $\mathbf{E}_{11}$, $\mathbf{E}_{12}$, $\mathbf{E}_{21}$ and $\mathbf{E}_{22}$:

$$\begin{aligned} \mathbf{G}_{11} &= \mathbf{A}_{11}\mathbf{S}_{11}\mathbf{A}_{11}^T + \mathbf{A}_{12}\mathbf{S}_{22}\mathbf{A}_{12}^T + \mathbf{O}_{11}\mathbf{S}_{11}\mathbf{A}_{11}^T + \mathbf{O}_{12}\mathbf{S}_{22}\mathbf{A}_{12}^T \\ \mathbf{G}_{12} &= \mathbf{A}_{11}\mathbf{S}_{11}\mathbf{A}_{21}^T + \mathbf{A}_{12}\mathbf{S}_{22}\mathbf{A}_{22}^T + \mathbf{O}_{11}\mathbf{S}_{11}\mathbf{A}_{21}^T + \mathbf{O}_{12}\mathbf{S}_{22}\mathbf{A}_{22}^T \\ \mathbf{G}_{21} &= \mathbf{A}_{21}\mathbf{S}_{11}\mathbf{A}_{11}^T + \mathbf{A}_{22}\mathbf{S}_{22}\mathbf{A}_{12}^T + \mathbf{O}_{21}\mathbf{S}_{11}\mathbf{A}_{11}^T + \mathbf{O}_{22}\mathbf{S}_{22}\mathbf{A}_{12}^T \\ \mathbf{G}_{22} &= \mathbf{A}_{21}\mathbf{S}_{11}\mathbf{A}_{21}^T + \mathbf{A}_{22}\mathbf{S}_{22}\mathbf{A}_{22}^T + \mathbf{O}_{21}\mathbf{S}_{22}\mathbf{A}_{21}^T + \mathbf{O}_{22}\mathbf{S}_{22}\mathbf{A}_{22}^T \end{aligned} \tag{4.3.28}$$

$$\begin{aligned} \mathbf{E}_{11} &= \mathbf{V}_{11} + \tfrac{1}{2}(\mathbf{A}_{11}\mathbf{S}_{11}\mathbf{O}_{11}^T + \mathbf{A}_{12}\mathbf{S}_{22}\mathbf{O}_{12}^T) + \tfrac{1}{2}(\mathbf{O}_{11}\mathbf{S}_{11}\mathbf{O}_{11}^T + \mathbf{O}_{12}\mathbf{S}_{22}\mathbf{O}_{12}^T) \\ \mathbf{E}_{12} &= \mathbf{V}_{12} + \tfrac{1}{2}(\mathbf{A}_{11}\mathbf{S}_{11}\mathbf{O}_{21}^T + \mathbf{A}_{12}\mathbf{S}_{22}\mathbf{O}_{22}^T) + \tfrac{1}{2}(\mathbf{O}_{11}\mathbf{S}_{11}\mathbf{O}_{21}^T + \mathbf{O}_{12}\mathbf{S}_{22}\mathbf{O}_{22}^T) \\ \mathbf{E}_{21} &= \mathbf{V}_{21} + \tfrac{1}{2}(\mathbf{A}_{21}\mathbf{S}_{11}\mathbf{O}_{11}^T + \mathbf{A}_{22}\mathbf{S}_{22}\mathbf{O}_{12}^T) + \tfrac{1}{2}(\mathbf{O}_{21}\mathbf{S}_{12}\mathbf{O}_{11}^T + \mathbf{O}_{22}\mathbf{S}_{22}\mathbf{O}_{12}^T) \\ \mathbf{E}_{22} &= \mathbf{V}_{22} + \tfrac{1}{2}(\mathbf{A}_{21}\mathbf{S}_{11}\mathbf{O}_{21}^T + \mathbf{A}_{22}\mathbf{S}_{22}\mathbf{O}_{22}^T) + \tfrac{1}{2}(\mathbf{O}_{21}\mathbf{S}_{11}\mathbf{O}_{21}^T + \mathbf{O}_{22}\mathbf{S}_{22}\mathbf{O}_{22}^T) . \end{aligned}$$

By using equation (4.3.28), equation (4.3.26) can be divided into two sets of equations. One takes the form

$$\mathbf{G}_{11}\mathbf{U}_s - \mathbf{E}_{11}\mathbf{W}^T\mathbf{U}_l + \mathbf{E}_{12}\mathbf{U}_l = \mathbf{Q}_1 \tag{4.3.29}$$

and the other

$$(\mathbf{G}_{21} - \mathbf{W}\mathbf{G}_{11})\mathbf{U}_s - (\mathbf{E}_{21} - \mathbf{W}\mathbf{E}_{11})\mathbf{W}^T\mathbf{U}_l + (\mathbf{E}_{22} - \mathbf{W}\mathbf{E}_{12})\mathbf{U}_l = \mathbf{Q}_2 - \mathbf{W}\mathbf{Q}_1 \; . \tag{4.3.30}$$

The nature of matrix $\mathbf{G}$ and equation (4.3.24) implies that the contribution of the first term in equation (4.3.30) can be ignored and equation (4.3.30) takes the form

$$-(\mathbf{E}_{21} - \mathbf{W}\mathbf{E}_{11})\mathbf{W}^T\mathbf{U}_l + (\mathbf{E}_{22} - \mathbf{W}\mathbf{E}_{12})\mathbf{U}_l = \mathbf{Q}_2 - \mathbf{W}\mathbf{Q}_1 \; . \tag{4.3.31}$$

Equation (4.3.29) can be used to determine $\mathbf{U}_s$ where $\mathbf{U}_l$ is known. In this case, $\mathbf{U}_s$ takes the form

$$\mathbf{U}_s = (\mathbf{G}_{11})^{-1}(\mathbf{Q}_1 + \mathbf{E}_{11}\mathbf{W}^T\mathbf{U}_l - \mathbf{E}_{12}\mathbf{U}_l) \; . \tag{4.3.32}$$

$\mathbf{U}_l$ is predicted by using equation (4.3.31). Because the elements of matrix $\mathbf{E}$ are a second-order function of the elements of vector $\mathbf{U}_l$, equation (4.3.30) is a set of simultaneous non-linear equations.

Such a set of equations is not easy to analyse. A possible method af analysis is to use a linear approximation. The solution obtained by linear approximation can be improved by successive approximations. For example, the first approximation of $\mathbf{U}_l$ denoted by $(\mathbf{U}_l)_1$ can be obtained by ignoring the higher-order effect in equation (4.3.31) altogether. In this case, $\mathbf{E}$ takes the form

$$\mathbf{E} = \mathbf{V} \; . \tag{4.3.33}$$

In this case, $\mathbf{V}$ is not a function of the elements of the vectors $\frac{1}{2}\mathbf{ASO}^T$ and $\frac{1}{2}\mathbf{OSO}^T$ all of which cause the higher-order effect of the nodal displacements in equation (4.3.31). By using equation (4.3.33) and equation (4.3.31), $(\mathbf{U}_l)_1$ takes the form

$$(\mathbf{U}_l)_1 = [(\mathbf{V}_{22} - \mathbf{W}\mathbf{V}_{12}) - (\mathbf{V}_{21} - \mathbf{W}\mathbf{V}_{11})\mathbf{W}^T]^{-1}(\mathbf{Q}_2 - \mathbf{W}\mathbf{Q}_1) \; . \tag{4.3.34}$$

$(\mathbf{U}_l)_1$, which is found by using linear approximation, can be improved by successive approximation.

The method of successive approximation is described here for the case of two non-linear equations, but it can easily be extended to cases with more unknowns. For two non-linear equations with two unknowns υ and w, the equations take the form

$$F(\upsilon, w) = 0\ , \qquad \mathrm{K}(\upsilon, w) = 0\ . \tag{4.3.35}$$

Assume that an approximate solution $(\upsilon)_1$, $(w)_1$ is known. An improved solution $(\upsilon)_2$, $(w)_2$ can be obtained. Let $(\upsilon)_2 - (\upsilon)_1 = \delta\upsilon$ and $(w)_2 - (w)_1 = \delta w$. By using Taylor's series, $F(\upsilon, w)$ and $K(\upsilon, w)$ can be expanded to terms of the first degree:

$$\begin{aligned} 0 &= F + F_\upsilon \delta_\upsilon + F_w \delta w \\ 0 &= K + K_\upsilon \delta_\upsilon + K_w \delta w \end{aligned} \tag{4.3.36}$$

where $F_\upsilon = \partial E/\partial\upsilon$ at $\upsilon = (\upsilon)_1$ and $w = (w)_1$, etc. The two equations can be used to determine $\delta\upsilon$ and δw. The new approximate solution is

$$\begin{aligned} (\upsilon)_2 &= (\upsilon)_1 + \delta\upsilon \\ (w)_2 &= (w)_1 + \delta w\ . \end{aligned} \tag{4.3.37}$$

The process can be repeated until the desired accuracy is obtained. For the cable net shown in Fig. 3.2.1(b), $\mathbf{U}_l^0$, $\mathbf{U}_l$, $\mathbf{U}_s^0$ and $\mathbf{U}_s$ take the forms

$$\mathbf{U}_l^0 = \begin{bmatrix} -U_l^8 \\ -U_l^8 \\ -U_l^8 \\ U_l^8 \\ U_l^8 \\ -U_l^8 \\ U_l^8 \\ U_l^8 \end{bmatrix}, \quad \mathbf{U}_l = [U_l^8]\ ; \quad \mathbf{U}_s^0 = \begin{bmatrix} U_s^1 \\ U_s^2 \\ U_s^3 \\ U_s^4 \\ U_s^5 \\ U_s^6 \\ U_s^7 \\ 0 \end{bmatrix}, \quad \mathbf{U}_s = \begin{bmatrix} U_s^1 \\ U_s^2 \\ U_s^3 \\ U_s^4 \\ U_s^5 \\ U_s^6 \\ U_s^7 \end{bmatrix}. \tag{4.3.38}$$

In this case, equation (4.3.31) takes the form of one third-order algebraic equation by which U_l^8 can be analysed easily. When $\mathbf{U}_l$ is analysed $\mathbf{U}_s$ can be determined by using equation (4.3.32) which is a linear set of seven equations with seven unknowns. Thus the non-linear effect of the large nodal displacements is not involved in solving a non-linear set of eight equations with eight unknowns but one non-linear equation and one set of seven linear equations. This is a tremendous reduction in the amount of work required to solve the problem.

For the tensegric shell shown in Fig. 3.2.2(b), $\mathbf{U}_l^0$, $\mathbf{U}_l$, $\mathbf{U}_s^0$ and $\mathbf{U}_s$ take the forms

$$\mathbf{U}_l^0 = \begin{bmatrix} \frac{2\sqrt{3}}{3} U_l^8 + U_l^9 \\ 0 \\ \frac{2\sqrt{3}}{3} U_l^8 + U_l^9 \\ 0 \\ \frac{\sqrt{3}}{3} U_l^8 + U_l^9 \\ U_l^8 \\ \frac{\sqrt{3}}{3} U_l^8 + U_l^9 \\ 0 \\ U_l^9 \\ 0 \\ U_l^8 \\ U_l^9 \end{bmatrix}, \quad \mathbf{U}_l = \begin{bmatrix} U_l^8 \\ U_l^9 \end{bmatrix}; \quad \mathbf{U}_s^0 = \begin{bmatrix} U_s^1 \\ U_s^2 \\ U_s^3 \\ U_s^4 \\ U_s^5 \\ U_s^6 \\ U_s^7 \\ U_s^{10} \\ U_s^{11} \\ U_s^{12} \\ 0 \\ 0 \end{bmatrix}, \quad \mathbf{U}_s = \begin{bmatrix} U_s^1 \\ U_s^2 \\ U_s^3 \\ U_s^4 \\ U_s^5 \\ U_s^6 \\ U_s^7 \\ U_s^{10} \\ U_s^{11} \\ U_s^{13} \end{bmatrix}. \tag{4.3.39}$$

In this case, equation (4.3.31) is a set of two third-order equations of U_l^8 and U_l^9. A first approximation of U_l^8 and U_l^9 can be obtained by using equation (4.3.34). The solution can be improved by using the method of successive approximation. After $\mathbf{U}_l$ is established, $\mathbf{U}_s$ can be analysed by using equation (4.3.32) which is a linear set of ten equations with ten unknowns. Also in this case the analysis of the problem involves solving a non-linear set of two equations only. The rest of the unknowns, ten unknowns, are analysed by using linear algebra.

For the cable net shown in Fig. 3.2.3, $\mathbf{U}_l^0$ and $\mathbf{U}_s^0$ take the forms

$$\mathbf{U}_l^0 = \begin{bmatrix} -0.749U_l^5 + 0.375U_l^{18} \\ -1.000U_l^5 + 0.500U_l^{18} \\ -2.000U_l^5 + 1.000U_l^{18} \\ -0.249U_l^5 + 0.249U_l^{18} \\ -0.500U_l^{18} \\ 2.000U_l^5 \\ 0.375U_l^{18} \\ -0.500U_l^{18} \\ -1.000U_l^{18} \\ 0.749U_l^5 + 0.375U_l^{18} \\ -1.000U_l^5 + 0.500U_l^{18} \\ 2.000U_l^5 - 1.000U_l^{18} \\ 0.249U_l^5 - 0.249U_l^{18} \\ 1.000U_l^5 \\ -2.000U_l^5 \\ -0.375U_l^{18} \\ 1.000U_l^5 \\ 1.000U_l^{18} \end{bmatrix}, \quad \mathbf{U}_l = \begin{bmatrix} U_5 \\ U_{18} \end{bmatrix}; \quad \mathbf{U}_s^0 = \begin{bmatrix} U_s^1 \\ U_s^2 \\ U_s^3 \\ U_s^4 \\ U_s^{17} \\ U_s^6 \\ U_s^7 \\ U_s^8 \\ U_s^9 \\ U_s^{10} \\ U_s^{11} \\ U_s^{12} \\ U_s^{13} \\ U_s^{14} \\ U_s^{15} \\ U_s^{16} \\ 0 \\ 0 \end{bmatrix}, \quad \mathbf{U}_s = \begin{bmatrix} U_s^1 \\ U_s^2 \\ U_s^3 \\ U_s^4 \\ U_s^{17} \\ U_s^6 \\ U_s^7 \\ U_s^8 \\ U_s^9 \\ U_s^{10} \\ U_s^{11} \\ U_s^{12} \\ U_s^{13} \\ U_s^{14} \\ U_s^{15} \\ U_s^{16} \end{bmatrix}. \tag{4.3.40}$$

In this case, equation (4.3.31) is a set of two third-order equations of U_l^5 and U_l^{18}. $\mathbf{U}_s$ can be analysed by using equation (4.3.32) which is a linear set of 16 equations with ten unknowns. Also in this case the cable net is analysed by using only two sets of non-linear equations. The rest of the unknowns, 16 unknowns, are analysed by using linear algebra.

5

The dynamical behaviour of cabled structures

5.1 A FULLY CONSTRAINED CABLED STRUCTURE: FREE VIBRATION

In the analysis of the dynamical behaviour of a cabled structure it is assumed that the mass of the structure can be seen as concentrated at the nodes. Where the structure vibrates dynamically the forces caused by the acceleration of the masses should be considered. If the acceleration in the x direction of the mass m_i, attached to node i, is $\ddot{U}_{xi}$, the force applied at this node in the x direction is

$$F_{xi} = m_i \ddot{U}_{xi} . \tag{5.1.1}$$

The two dots on $\ddot{U}_{xi}$ indicate that the acceleration is the second derivative of the displacement with respect to time. Where the masses at all the nodes of the structure are considered, the equation takes the form

$$\mathbf{DU} + \mathbf{M\ddot{U}} = \mathbf{Q} , \tag{5.1.2}$$

M being the matrix of the appropriate masses. For a fully constrained cabled structure, matrix **D** is a square matrix in which all rows are independent and matrix **M** is diagonal.

For free vibration where no load is applied to the structure, equation (5.1.2) takes the form

$$\mathbf{DU} + \mathbf{M\ddot{U}} = 0 . \tag{5.1.3}$$

It is assumed that, in the free-vibration case, the nodal displacements follow a harmonic function:

$$\mathbf{U}_j = \mathbf{U}_{Mj} \sin(\omega_j t + \phi_j) \ . \tag{5.1.4}$$

$\mathbf{U}_j$ and $\mathbf{U}_{Mj}$ are the vectors of the nodal displacements and the vectors of the amplitude of the nodal displacement associated with mode j. ω_j and ϕ_j are the **angular frequency** and the **phase angle** of mode j. By using equation (5.1.4), equation (5.1.3) takes the form

$$\begin{aligned} &\mathbf{H}_j\mathbf{U}_{Mj} = 0 \\ &\mathbf{H}_j = \mathbf{D}(\omega_j)^2\mathbf{M} \ . \end{aligned} \tag{5.1.5}$$

Equation (5.1.5) is called a **self-adjoint equation.** $\mathbf{U}_{Mj}$ and ω_j which satisfy this equation are called **eigenfunctions** and **eigenvalues**. ω_j is the frequency of the vibration of the mode whose shape is given by the eigenfunction. Equation (5.1.5) is satisfied where

$$\mathbf{H}_j = 0 \ . \tag{5.1.6}$$

Equation (5.1.6) is satisfied where the determinant of $\mathbf{H}_j$ is equal to zero:

$$\det|\mathbf{H}_j| = 0 \ . \tag{5.1.7}$$

Expansion of this determinant yields a polynomial of $(\omega_j)^2$ in which the highest order is $(\omega_j)^{2n}$, where n is the number of rows or columns of matrix $\mathbf{D}$. The roots $(\omega_1)^2$, $(\omega_2)^2$, $(\omega_3)^2, \ldots$ of this polynomial can be found by using a numerical procedure. Where the value of $(\omega_j)^2$ is known, it is possible to use equation (5.1.5) to determine $\mathbf{U}_{Mj}$. Because of the nature of the equation, $n-1$ elements of vector $\mathbf{U}_{Mj}$ can be determined as a function of the remaining one term. It is possible to define $\mathbf{U}_{Mj}$ by considering the inverse of matrix $\mathbf{H}_j$. Because $\det|\mathbf{H}_j|$ is equal to zero, $(\mathbf{H}_j)^{-1}$ does not exist, but it is possible to write it in the following formal way:

$$(\mathbf{H}_j)^{-1} = \frac{\mathbf{H}_j^a}{\det|\mathbf{H}_j|} \tag{5.1.8}$$

where $\mathbf{H}_j^a$ is the adjoint matrix of $\mathbf{H}_j$, which is the transpose of the cofactor matrix $\mathbf{H}_j^c$:

$$\mathbf{H}_j^a = (\mathbf{H}_j^c)^T \ . \tag{5.1.9}$$

Equation (5.1.8) can be rewritten to take the form

$$\mathbf{H}_j\mathbf{H}_j^{\mathrm{a}} = \det|\mathbf{H}_j|\ \mathbf{I} = 0\ . \tag{5.1.10}$$

I is a unit matrix. Comparing equation (5.1.5) with (5.1.10) shows that $\mathbf{U}_{\mathrm{M}j}$ is proportional to any non-zero column of the adjoint matrix $\mathbf{H}_j^{\mathrm{a}}$.

By considering mode $\mathbf{U}_{\mathrm{M}j}$ and $\mathbf{U}_{\mathrm{M}k}$, equation (5.1.5) takes the form

$$\begin{aligned}\mathbf{D}\mathbf{U}_{\mathrm{M}j} &= (\omega_j)^2\mathbf{M}\mathbf{U}_{\mathrm{M}j}\\ \mathbf{D}\mathbf{U}_{\mathrm{M}k} &= (\omega_k)^2\mathbf{M}\mathbf{U}_{\mathrm{M}k}\ .\end{aligned} \tag{5.1.11}$$

Because matrix **M** is diagonal and matrix **D** is symmetrical, the transpose of the second equation of equation (5.1.11) takes the form

$$(\mathbf{U}_{\mathrm{M}k})^{\mathrm{T}}\mathbf{D} = (\omega_k)^2(\mathbf{U}_{\mathrm{M}k})^{\mathrm{T}}\mathbf{M}\ . \tag{5.1.12}$$

By multiplication of the first equation of equation (5.1.11) by $(\mathbf{U}_{\mathrm{M}k})^{\mathrm{T}}$ and equation (5.1.12) by $\mathbf{U}_{\mathrm{M}j}$, appropriately, they take the forms

$$\begin{aligned}(\mathbf{U}_{\mathrm{M}k})^{\mathrm{T}}\mathbf{D}\mathbf{U}_{\mathrm{M}j} &= (\omega_j)^2(\mathbf{U}_{\mathrm{M}k})^{\mathrm{T}}\mathbf{M}\mathbf{U}_{\mathrm{M}j}\\ (\mathbf{U}_{\mathrm{M}k})^{\mathrm{T}}\mathbf{D}\mathbf{U}_{\mathrm{M}j} &= (\omega_k)^2(\mathbf{U}_{\mathrm{M}k})^{\mathrm{T}}\mathbf{M}\mathbf{U}_{\mathrm{M}j}\ .\end{aligned} \tag{5.1.13}$$

Because the left-hand sides of the equations in equation (5.1.13) are equal,

$$[(\omega_j)^2 - (\omega_k)^2](\mathbf{U}_{\mathrm{M}k})^{\mathrm{T}}\mathbf{M}\mathbf{U}_{\mathrm{M}j} = 0\ . \tag{5.1.14}$$

Dividing the first equation of equation (5.1.13) by $(\omega_j)^2$ and the second by $(\omega_k)^2$ and considering the fact that the right-hand sides of the two equations are the same,

$$\left(\frac{1}{(\omega_j)^2} - \frac{1}{(\omega_k)^2}\right)(\mathbf{U}_{\mathrm{M}k})^{\mathrm{T}}\mathbf{D}\mathbf{U}_{\mathrm{M}j} = 0\ . \tag{5.1.15}$$

Equations (5.1.14) and (5.1.15) imply that

$$(\mathbf{U}_{\mathrm{M}k})^{\mathrm{T}}\mathbf{M}\mathbf{U}_{\mathrm{M}j} = 0,\ (\mathbf{U}_{\mathrm{M}k})^{\mathrm{T}}\mathbf{D}\mathbf{U}_{\mathrm{M}j} = 0\ . \tag{5.1.16}$$

Equation (5.1.16) indicates the **orthogonality** of the modes of vibration.

Where $\omega_j = \omega_k$, it is a possible to define $\mathbf{U}_{\mathrm{M}j}$ so that

$$(\mathbf{U}_{\mathrm{M}j})^{\mathrm{T}}\mathbf{M}\mathbf{U}_{\mathrm{M}j} = \mathbf{I}\ . \tag{5.1.17}$$

When $\mathbf{U}_{Mj}$ is arranged to satisfy equation (5.1.17), it is considered that $\mathbf{U}_{Mj}$ is normalized with respect to the mass matrix. In this case it is denoted by $\mathbf{U}_{Nj}$. By using $\mathbf{U}_{Nj}$, equation (5.1.11) takes the form

$$(\mathbf{U}_{Nj})^{T}\mathbf{D}\mathbf{U}_{Nj} = (\omega_j)^2 \ . \tag{5.1.18}$$

When the modes are normalized with respect to **M, D** is equal to the square of the appropriate frequency.

For convenience it is possible to arrange vectors $\mathbf{U}_{Nj}$ in a matrix form $\mathbf{U}_N$:

$$\mathbf{U}_N = [\mathbf{U}_{N1}\ \mathbf{U}_{N2}\ \mathbf{U}_{N3} \ldots] \ . \tag{5.1.19}$$

For example, for the cable net shown in Fig. 3.2.1(a), **A** takes the form

$$\mathbf{A} = \begin{bmatrix} \frac{\sqrt{3}}{2} & -\frac{\sqrt{3}}{2} & \frac{1}{2} & -\frac{1}{2} \\ -\frac{1}{2} & -\frac{1}{2} & \frac{\sqrt{3}}{2} & \frac{\sqrt{3}}{2} \end{bmatrix} \ . \tag{5.1.20}$$

In the case where the cross-section of all cables is A_0 and they are made of the same material, **S** takes the form

$$\mathbf{S} = \frac{EA_0}{a}\begin{bmatrix} \sqrt{3} & 0 & 0 & 0 \\ 0 & \sqrt{3} & 0 & 0 \\ 0 & 0 & 1 & 0 \\ 0 & 0 & 0 & 1 \end{bmatrix} \ . \tag{5.1.21}$$

In the case where the mass m_0 is concentrated at node 1, **M** takes the form

$$\mathbf{M} = m_0\begin{bmatrix} 1 & 0 \\ 0 & 1 \end{bmatrix} . \tag{5.1.22}$$

By using equations (5.1.20)–(5.1.22), equation (5.1.3) takes the form

$$\frac{EA_0}{a}\begin{bmatrix}\dfrac{3\sqrt{3}}{2} & 0\\ 0 & \dfrac{\sqrt{3}+3}{2}\end{bmatrix}\mathbf{U}+m_0\begin{bmatrix}1 & 0\\ 0 & 1\end{bmatrix}\ddot{\mathbf{U}}=0\ . \tag{5.1.23}$$

Matrix $\mathbf{H}_j$ takes the form

$$\mathbf{H}_j=\begin{bmatrix}\dfrac{3\sqrt{3}EA_0}{2a}-(\omega_j)^2m_0 & 0\\ 0 & \dfrac{(\sqrt{3}+3)EA_0}{2a}-(\omega_j)^2m_0\end{bmatrix}. \tag{5.1.24}$$

$\det|\mathbf{H}_j|$ takes the form

$$\left(\frac{3\sqrt{3}EA_0}{2a}-(\omega_j)^2m_0\right)\left(\frac{(3+3)EA_0}{2a}(\omega_j)^2m_0\right)=0\ . \tag{5.1.25}$$

It can be seen that equation (5.1.25) is satisfied, where

$$\frac{3\sqrt{3}EA_0}{2a}-(\omega_j)^2m_0=0 \tag{5.1.26}$$

or where

$$\frac{(\sqrt{3}+3)EA_0}{2a}-(\omega_j)^2m_0=0\ . \tag{5.1.27}$$

Equation (5.1.26) implies that one frequency of free vibration is

$$\omega_1=\sqrt{\frac{3\sqrt{3}EA_0}{2am_0}} \tag{5.1.28}$$

and equation (5.1.27) implies that the second frequency of free vibration is

$$\omega_2 = \sqrt{\frac{(\sqrt{3}+3)EA_0}{2am_0}} . \tag{5.1.29}$$

In order to determine the modes of free vibration, the transposed matrix of the cofactors of $\mathbf{H}_j$ is investigated. The mode associated with the first frequency is investigated by using $(\mathbf{H}_j^c)^T$ where $\omega_j = \omega_i$. In this case, $(\mathbf{H}_j^c)^T$ takes the form

$$(\mathbf{H}_1^c)^T = \begin{bmatrix} \frac{(3-2\sqrt{3})EA_0}{3a} & 0 \\ 0 & 0 \end{bmatrix} . \tag{5.1.30}$$

It can be seen that the first column only can be used and $\mathbf{U}_{M1}$ takes the form

$$\mathbf{U}_{M1} = U_{M1}^0 \begin{bmatrix} 1 \\ 0 \end{bmatrix} \tag{5.1.31}$$

where U_{M1}^0 is a constant to be determined. By using the same method for ω_2, $\mathbf{U}_{M2}$ takes the form

$$\mathbf{U}_{M2} = U_{M2}^0 \begin{bmatrix} 0 \\ 1 \end{bmatrix} \tag{5.1.32}$$

where $\mathbf{U}_{M1}$ and $\mathbf{U}_{M2}$ are normalized with respect to

$$m_0\, U_{M1}^0\, [1 \quad 0] \begin{bmatrix} 1 & 0 \\ 0 & 1 \end{bmatrix} \begin{bmatrix} 1 \\ 0 \end{bmatrix} = 1 \tag{5.1.33}$$

which implies that

$$U_{M1}^0 = \frac{1}{m_0} . \tag{5.1.34}$$

In the same way it is found that

$$U^0_{M2} = \frac{1}{m_0} . \tag{5.1.35}$$

By using equations (5.1.31), (5.1.32), (5.1.34) and (5.1.35), equation (5.1.19) takes the form

$$\mathbf{U}_N = \frac{1}{m_0}\begin{bmatrix} 1 & 0 \\ 0 & 1 \end{bmatrix} . \tag{5.1.36}$$

5.2 A FULLY CONSTRAINED CABLED STRUCTURE: FORCED VIBRATION

In the case where the structure is loaded by a dynamical load $\mathbf{Q}$, the structure response is given by equation (5.1.2). By multiplying both sides with $(\mathbf{U}_N)^T$, equation (5.1.2) takes the form

$$(\mathbf{U}_N)^T\mathbf{D}\mathbf{U} + (\mathbf{U}_N)^T\mathbf{M}\ddot{\mathbf{U}} = (\mathbf{U}_N)^T\mathbf{Q} . \tag{5.2.1}$$

It is possible to define the **principal nodal displacements** $\mathbf{U}_P$:

$$\begin{aligned} \mathbf{U} &= \mathbf{U}_N\mathbf{U}_P \\ \ddot{\mathbf{U}} &= \mathbf{U}_N\ddot{\mathbf{U}}_P . \end{aligned} \tag{5.2.2}$$

By using equation (5.2.2), transformation from $\mathbf{U}$ to $\mathbf{U}_P$ takes the form

$$\begin{aligned} \mathbf{U}_P &= (\mathbf{U}_N)^{-1}\mathbf{U} \\ \ddot{\mathbf{U}}_P &= (\mathbf{U}_N)^{-1}\ddot{\mathbf{U}} . \end{aligned} \tag{5.2.3}$$

By using equation (5.2.2), equation (5.2.1) takes the form

$$(\mathbf{U}_N)^T\mathbf{D}\mathbf{U}_N\mathbf{U}_P + (\mathbf{U}_N)^T\mathbf{M}\mathbf{U}_N\ddot{\mathbf{U}}_P = (\mathbf{U}_N)^T\mathbf{Q} . \tag{5.2.4}$$

Equation j of equation (5.2.4) takes the form

$$(\omega_j)^2U_{Pj} + \ddot{U}_{Pj} = Q_{Pj} \tag{5.2.5}$$

in which Q_{Pj} takes the form

$$Q_{Pj} = U_{N1j}Q_1 + U_{N2j}Q_2 + U_{N3j}Q_3 + \dots . \tag{5.2.6}$$

In the case where $Q_{\mathrm{P}j} = 0$, the solution of equation (5.2.5) takes the form

$$U_{\mathrm{P}j} = U^{\mathrm{c}}_{\mathrm{P}j} \cos(\omega_j t) + \frac{U^{0}_{\mathrm{P}j}}{\omega_j} \sin(\omega_j t) \tag{5.2.7}$$

where $U^{0}_{\mathrm{P}j}$ and $\dot{U}^{0}_{\mathrm{P}j}$ indicate the displacement and velocity, respectively, at $t = 0$.

In the case where the load $Q_{\mathrm{P}j}$ is a function of time the load applied between t' and $t' + \mathrm{d}t'$ can be seen as an impulse of $Q_{\mathrm{P}j}\,\mathrm{d}t'$. The impulse causes an increase in velocity given by

$$\mathrm{d}\dot{U}_{\mathrm{P}j} = Q_{\mathrm{P}j}\,\mathrm{d}t' \; . \tag{5.2.8}$$

By using equation (5.2.7), the increase in displacement due to an increase in velocity at t' takes the form

$$\mathrm{d}U_{\mathrm{P}j} = \frac{Q_{\mathrm{P}j}}{\omega_j} \sin[\omega_j(t - t')]\,\mathrm{d}t' \tag{5.2.9}$$

at the later time t. The result of a continuous action from 0 to t takes the form

$$U_{\mathrm{P}j} = \frac{1}{\omega_j} \int_0^t Q_{\mathrm{P}j} \sin[\omega_j(t - t')]\,\mathrm{d}t' \; . \tag{5.2.10}$$

Equation (5.2.10) is known as Duhamel's integral. To take account of $U^{0}_{\mathrm{P}j}$ and $\dot{U}^{0}_{\mathrm{P}j}$, equation (5.2.7) should be added to (5.2.10). The general solution is

$$U_{\mathrm{P}j} = U^{0}_{\mathrm{P}j} \cos(\omega_j t) + \frac{\dot{U}^{0}_{\mathrm{P}j}}{\omega_j} \sin(\omega_j t) + \frac{1}{\omega_j} \int_0^t Q_{\mathrm{P}j} \sin[\omega_j(t - t')]\,\mathrm{d}t' \; . \tag{5.2.11}$$

Equation (5.2.11) can be used to determine the principal nodal displacements, considering the dynamical load applied to a fully constrained cabled structure. By using equation (5.2.2), it is easy to transfer the solution to the actual nodal displacements.

For the cable net shown in Fig. 3.2.1(a),

$$\begin{aligned} \mathbf{U}_{\mathrm{P}} &= \mathbf{U} \\ \ddot{\mathbf{U}}_{\mathrm{P}} &= \ddot{\mathbf{U}} \; . \end{aligned} \tag{5.2.12}$$

Where the net load takes the form

$$\mathbf{Q} = \begin{bmatrix} Q_x \\ Q_y \end{bmatrix} \tag{5.2.13}$$

in which Q_x and Q_y are constant loads suddenly applied to the mass at time $t=0$. equation (5.2.4) takes the form

$$\begin{aligned} (\omega_1)^2 U_1 + \ddot{U}_1 &= \frac{Q_x}{m_0} \\ (\omega_2)^2 U_2 + \ddot{U}_2 &= \frac{Q_y}{m_0} \ . \end{aligned} \tag{5.2.14}$$

By using Duhamel's integral,

$$\begin{aligned} U_1 &= \frac{Q_x}{m_0\omega_1} \int_0^t \sin[\omega_1(t-t')]\ \mathrm{d}t' = \frac{Q_x}{m_0(\omega_1)^2}\,[1-\cos(\omega_1 t)] \\ U_2 &= \frac{Q_x}{m_0\omega_2} \int_0^t \sin[\omega_2(t-t')]\ \mathrm{d}t' = \frac{Q_x}{m_0(\omega_1)^2}\,[1-\cos(\omega_2 t)]\ . \end{aligned} \tag{5.2.15}$$

By using equation (5.2.14) it can be seen that, where there is no dynamical effect, i.e. $\ddot{U}_1 = \ddot{U}_2 = 0$, the nodal displacements denoted by $(U_1)_0$ and $(U_2)_0$ take the form

$$\begin{aligned} (U_1)_0 &= \frac{Q_x}{m_0(\omega_1)^2} \\ (U_2)_0 &= \frac{Q_y}{m_0(\omega_2)^2}\ . \end{aligned} \tag{5.2.16}$$

By using equation (5.2.16), equation (5.2.15) takes the form

$$\begin{aligned} U_1 &= (U_1)_0[1-\cos(\omega_1 t)] \\ U_2 &= (U_2)_0[1-\cos(\omega_2 t)]. \end{aligned} \tag{5.2.17}$$

Equation (5.2.17) indicates that, because of the dynamical effect where load $\mathbf{Q}$ is applied suddenly, the amplitude of the maximum displacement is double the amplitude due to static loading.

5.3 AN UNDER-CONSTRAINED CABLED STRUCTURE: FREE VIBRATION

For an under-constrained cable net the response of the structure because of static load is given by equation (4.3.13). Where the dynamical effect of the vibration of the masses is considered, equation (4.3.13) takes the form

$$\mathbf{DU} + \mathbf{VU} + \tfrac{1}{2}\mathbf{ASO}^{\mathrm{T}}\mathbf{U} + \mathbf{OSA}^{\mathrm{T}}\mathbf{U} + \tfrac{1}{2}\mathbf{OSO}^{\mathrm{T}}\mathbf{U} + \mathbf{M}\ddot{\mathbf{U}} = \mathbf{Q} \tag{5.3.1}$$

where matrix **M** is a diagonal matrix and it is arranged in accordance with matrix **D** and vector **U**. Element ii of matrix **M** denoted by m_{ii} is the mass associated with acceleration in the U_i direction. For the cable net shown in Fig. 3.2.1(b) where vector **U** takes the form given by equation (3.4.14), matrix **M** takes the form

$$\mathbf{M} = \left[\begin{array}{ccccccc|c} m_1 & 0 & 0 & 0 & 0 & 0 & 0 & 0 \\ 0 & m_1 & 0 & 0 & 0 & 0 & 0 & 0 \\ 0 & 0 & m_2 & 0 & 0 & 0 & 0 & 0 \\ 0 & 0 & 0 & m_2 & 0 & 0 & 0 & 0 \\ 0 & 0 & 0 & 0 & m_3 & 0 & 0 & 0 \\ 0 & 0 & 0 & 0 & 0 & m_3 & 0 & 0 \\ 0 & 0 & 0 & 0 & 0 & 0 & m_4 & 0 \\ \hline 0 & 0 & 0 & 0 & 0 & 0 & 0 & m_4 \end{array}\right] \quad \begin{array}{l} \mathbf{M}_{11},\ \mathbf{M}_{12} \\ \mathbf{M}_{21},\ \mathbf{M}_{22} \end{array} \tag{5.3.2}$$

m_i indicates the mass attached to node i. For the tensegric shell shown in Fig. 3.2.2(b), where vector **U** takes the form given by equation (3.4.36), matrix **M** takes the form

$$\mathbf{M} = \left[\begin{array}{cccccccccc|cc} m_1 & 0 & 0 & 0 & 0 & 0 & 0 & 0 & 0 & 0 & 0 & 0 \\ 0 & m_1 & 0 & 0 & 0 & 0 & 0 & 0 & 0 & 0 & 0 & 0 \\ 0 & 0 & m_2 & 0 & 0 & 0 & 0 & 0 & 0 & 0 & 0 & 0 \\ 0 & 0 & 0 & m_2 & 0 & 0 & 0 & 0 & 0 & 0 & 0 & 0 \\ 0 & 0 & 0 & 0 & m_3 & 0 & 0 & 0 & 0 & 0 & 0 & 0 \\ 0 & 0 & 0 & 0 & 0 & m_3 & 0 & 0 & 0 & 0 & 0 & 0 \\ 0 & 0 & 0 & 0 & 0 & 0 & m_4 & 0 & 0 & 0 & 0 & 0 \\ 0 & 0 & 0 & 0 & 0 & 0 & 0 & m_5 & 0 & 0 & 0 & 0 \\ 0 & 0 & 0 & 0 & 0 & 0 & 0 & 0 & m_6 & 0 & 0 & 0 \\ 0 & 0 & 0 & 0 & 0 & 0 & 0 & 0 & 0 & m_6 & 0 & 0 \\ \hline 0 & 0 & 0 & 0 & 0 & 0 & 0 & 0 & 0 & 0 & m_4 & 0 \\ 0 & 0 & 0 & 0 & 0 & 0 & 0 & 0 & 0 & 0 & 0 & m_5 \end{array}\right] \quad \begin{array}{l} \mathbf{M}_{11},\ \mathbf{M}_{12} \\ \mathbf{M}_{21},\ \mathbf{M}_{22} \end{array} \tag{5.3.3}$$

For the cable net shown in Fig. 3.2.3 where vector **U** takes the form given by equation (3.4.37), matrix **M** takes the form

$$\mathbf{M}=\left[\begin{array}{cccccccccccccccc|cc}
\multicolumn{16}{c|}{\mathbf{M}_{11}} & \multicolumn{2}{c}{\mathbf{M}_{12}} \\
m_1&0&0&0&0&0&0&0&0&0&0&0&0&0&0&0&0&0\\
0&m_1&0&0&0&0&0&0&0&0&0&0&0&0&0&0&0&0\\
0&0&m_1&0&0&0&0&0&0&0&0&0&0&0&0&0&0&0\\
0&0&0&m_2&0&0&0&0&0&0&0&0&0&0&0&0&0&0\\
0&0&0&0&m_6&0&0&0&0&0&0&0&0&0&0&0&0&0\\
0&0&0&0&0&m_2&0&0&0&0&0&0&0&0&0&0&0&0\\
0&0&0&0&0&0&m_3&0&0&0&0&0&0&0&0&0&0&0\\
0&0&0&0&0&0&0&m_3&0&0&0&0&0&0&0&0&0&0\\
0&0&0&0&0&0&0&0&m_3&0&0&0&0&0&0&0&0&0\\
0&0&0&0&0&0&0&0&0&m_4&0&0&0&0&0&0&0&0\\
0&0&0&0&0&0&0&0&0&0&m_4&0&0&0&0&0&0&0\\
0&0&0&0&0&0&0&0&0&0&0&m_4&0&0&0&0&0&0\\
0&0&0&0&0&0&0&0&0&0&0&0&m_5&0&0&0&0&0\\
0&0&0&0&0&0&0&0&0&0&0&0&0&m_5&0&0&0&0\\
0&0&0&0&0&0&0&0&0&0&0&0&0&0&m_5&0&0&0\\
0&0&0&0&0&0&0&0&0&0&0&0&0&0&0&m_6&0&0\\
\hline
0&0&0&0&0&0&0&0&0&0&0&0&0&0&0&0&m_2&0\\
0&0&0&0&0&0&0&0&0&0&0&0&0&0&0&0&0&m_6\\
\multicolumn{16}{c|}{\mathbf{M}_{21}} & \multicolumn{2}{c}{\mathbf{M}_{22}}
\end{array}\right] . \tag{5.3.4}$$

For free vibration, where small nodal displacements are considered, the second and third order of the displacements can be ignored and equation (5.3.1) takes the form

$$\mathbf{DU}+\mathbf{VU}+\mathbf{M\ddot{U}}=0 . \tag{5.3.5}$$

It is assumed that the nodal displacements in free vibration follow the harmonic function given by equation (5.1.4). By using equation (5.1.4), equation (5.3.5) takes the form

$$\mathbf{DU}_{\mathrm{M}j}+\mathbf{VU}_{\mathrm{M}j}-(\omega_j)^2\mathbf{UV}_{\mathrm{M}j}=0 . \tag{5.3.6}$$

$\mathbf{U}_{\mathrm{M}j}$ can be partitioned into $\mathbf{U}_{\mathrm{M}1j}$ and $\mathbf{U}_{\mathrm{M}2j}$ in accordance with $\mathbf{U}_1$ and $\mathbf{U}_2$. Matrix **M** can be partitioned in accordance with $\mathbf{U}_1$, $\mathbf{U}_2$, $\mathbf{D}_{11}$, $\mathbf{D}_{12}$, $\mathbf{D}_{21}$ and $\mathbf{D}_{22}$. Because matrix **M** is diagonal, $\mathbf{M}_{12}=0$ and $\mathbf{M}_{21}=0$. $\mathbf{M}_{11}$, $\mathbf{M}_{12}$, $\mathbf{M}_{21}$ and $\mathbf{M}_{22}$ are given by equation (5.3.2) for the cable net shown in Fig. 3.2.1(b), by equation (5.3.3) for the tensegric shell shown in Fig. 3.2.2(b), and by equation (5.3.4) for the cable net shown in Fig. 3.2.3. By using $\mathbf{U}_{\mathrm{M}1j}$ and $\mathbf{U}_{\mathrm{M}2j}$, equation (5.3.6) can be divided into two sets of equations. One set takes the form

$$\mathbf{D}_{11}\mathbf{U}_{\mathrm{M}1j}+\mathbf{D}_{12}\mathbf{U}_{\mathrm{M}2j}+\mathbf{V}_{11}\mathbf{U}_{\mathrm{M}1j}+\mathbf{V}_{12}\mathbf{U}_{\mathrm{M}2j}-(\omega_j)^2\mathbf{M}_{11}\mathbf{U}_{\mathrm{M}1j}=0 \tag{5.3.7}$$

and the second set takes the form

$$(\mathbf{V}_{21}-\mathbf{W}\mathbf{V}_{11})\mathbf{U}_{M1j}+(\mathbf{V}_{22}-\mathbf{W}\mathbf{V}_{12})\mathbf{U}_{M2j}-(\omega_j)^2(-\mathbf{W}\mathbf{M}_{11}\ \mathbf{U}_{M1j}+\mathbf{M}_{22}\mathbf{U}_{M2j})=0. \tag{5.3.8}$$

In practical cases of small nodal displacements, the contribution of the third and fourth terms of equation (5.3.7) can be ignored. In this case, equation (5.3.7) takes the form

$$\mathbf{D}_{11}\mathbf{U}_{M1j}+\mathbf{D}_{12}\mathbf{U}_{M2j}-(\omega_j)^2\ \mathbf{M}_{11}\mathbf{U}_{M1j}=0\ . \tag{5.3.9}$$

It is possible to distinguish between two cases in the solution of equations (5.3.9) and (5.3.8). The case of low frequencies denoted by ω_j^a with the appropriate modes of displacements $\mathbf{U}_{Mj}^a$, and the case of high frequencies denoted by ω_j^g and the appropriate modes of displacements $\mathbf{U}_{Mj}^g$. In the case of low frequencies, the contribution of the third term of equation (5.3.9) can be ignored and equation (5.3.9) takes the form

$$\mathbf{D}_{11}\mathbf{U}_{M1j}^a+\mathbf{D}_{12}\mathbf{U}_{M2j}^a=0\ . \tag{5.3.10}$$

Because $\mathbf{D}_{12}=\mathbf{D}_{11}\mathbf{W}^T$, equation (5.3.10) is satisfied where

$$\mathbf{U}_{M1j}^a=-\mathbf{W}^T\ \mathbf{U}_{M2j}^a\ . \tag{5.3.11}$$

Equation (5.3.11) is similar to equation (4.3.17) which indicates that the nodal displacements associated with low frequencies of vibration are the nodal displacements caused by rigid-body movement of the members of the structure. By using equation (5.3.11), equation (5.3.8) takes the form

$$[(\mathbf{V}_{22}-\mathbf{W}\mathbf{V}_{12})-(\mathbf{V}_{21}-\mathbf{W}\mathbf{V}_{11})\mathbf{W}^T]\mathbf{U}_{M2j}^a-(\omega_j^a)^2(\mathbf{W}\mathbf{M}_{11}\mathbf{W}^T+\mathbf{M}_{22})\mathbf{U}_{M2j}^a=0\ . \tag{5.3.12}$$

Equation (5.3.12) is a self-adjoint equation which can be solved by using the methods discussed in section 5.1. By using these methods, $m-r$ different frequencies ω^a, where $m-r$ is the order of the vector $\mathbf{U}_{M2j}^a$ and associated modes $\mathbf{U}_{M2j}^a$, can be determined. By using equations (5.3.11) and (5.3.12), modes $\mathbf{U}_{Mj}^a$ associated with low frequencies take the form

$$\mathbf{U}_{Mj}^a=\begin{bmatrix}-\mathbf{W}^T\mathbf{U}_{M2j}^a\\ \mathbf{U}_{M2}^a\end{bmatrix}\ . \tag{5.3.13}$$

For the cable net shown in Fig. 3.2.1(b) there is one low-frequency mode of vibration. It is given by equation (4.3.18) and is shown in Fig. 4.3.2. For the tensegric shell shown in Fig. 3.2.2(b) there are two low-frequency modes of vibration. They are given by equation (4.3.19). For the cable net shown in Fig. 3.2.3 there are two low-frequency modes of vibration. They are given by equation (4.3.20) and are shown in Figs 4.3.3 and 4.3.4. By using these modes of vibration and by using equation (5.3.12), the mass of the net is assumed to be 32 N s^2 cm^{-1} concentrated at each node; the low frequencies of vibration are found to be

$$\begin{aligned} \omega_1^a &= 2.024 \text{ s}^{-1} \\ \omega_2^a &= 2.446 \text{ s}^{-1} . \end{aligned} \tag{5.3.14}$$

The low-frequency mode of vibration can be scaled arbitrarily. When it is normalized with respect to the mass matrix

$$(\mathbf{U}_{Nj}^a)^T \mathbf{M} \mathbf{U}_{Nj}^a = 1 \tag{5.3.15}$$

it takes the form

$$\mathbf{U}_{Nj}^a = \begin{bmatrix} -\mathbf{W}^T \mathbf{U}_{N2j}^a \\ \mathbf{U}_{N2j}^a \end{bmatrix} . \tag{5.3.16}$$

In the case of high frequencies the contribution of the first two terms of equation (5.3.8) can be ignored and equation (5.3.8) takes the form

$$\begin{aligned} \mathbf{U}_{M2j}^g &= \mathbf{N} \mathbf{U}_{Mij}^g \\ \mathbf{N} &= (\mathbf{M}_{22})^{-1} \mathbf{W} \mathbf{M}_{11} . \end{aligned} \tag{5.3.17}$$

By using equation (5.3.17), equation (5.3.9) takes the form

$$(\mathbf{D}_{11} + \mathbf{D}_{12}\mathbf{N}) \mathbf{U}_{M1j}^g - (\omega_j^g)^2 \mathbf{M}_{11} \mathbf{U}_{M1j}^g = 0 . \tag{5.3.18}$$

Equation (5.3.18) is a self-adjoint equation. By using it, the high frequencies and modes of vibration can be determined. In this case the number of different ω_j^g and $\mathbf{U}_{Mj}^g$ is equal to the order of the vector $\mathbf{U}_{M1j}^g$, i.e. r. For the cable net shown in Fig. 3.2.3 there are 16 high-frequency modes of vibration. Equation (5.3.18) is composed of 16 self-adjointed equations associated with 16 different high frequencies. The lowest of the high frequencies is found to be

$$\omega_1^g = 7.468 \text{ s}^{-1} . \tag{5.3.19}$$

Comparison of $(\omega_1^g)^2$ with $(\omega_2^a)^2$ (i.e. 5.983 with 55.779) indicates the large difference between the magnitude of the square of the frequencies of vibration at low and high frequencies. This result is typical of all under-constrained cabled structures. By using equations (5.3.17) and (5.3.18), the modes associated with the high frequencies take the form

$$\mathbf{U}_{Mj}^{g} = \begin{bmatrix} \mathbf{U}_{M1j}^{g} \\ \mathbf{N}\mathbf{U}_{M1j}^{g} \end{bmatrix} . \tag{5.3.20}$$

Where the mode is normalized with respect to the mass matrix, it takes the form

$$\mathbf{U}_{Nj}^{g} = \begin{bmatrix} \mathbf{U}_{N1j}^{g} \\ \mathbf{N}\mathbf{U}_{N1j}^{g} \end{bmatrix} . \tag{5.3.21}$$

By using equations (5.3.16) and (5.3.21), the free vibration takes the form:

$$\mathbf{U} = \sum_{f}^{m-r} \begin{bmatrix} -\mathbf{W}^{\mathrm{T}}\mathbf{U}_{N2j}^{a} \\ \mathbf{U}_{N2j}^{a} \end{bmatrix} \sin\,(\omega_j^a t + \phi_j) + \sum_{j}^{r} \begin{bmatrix} \mathbf{U}_{N1j}^{g} \\ \mathbf{N}\mathbf{U}_{N1j}^{g} \end{bmatrix} \sin(\omega_j^g t + \phi_j^g) \tag{5.3.22}$$

where $\mathbf{U}_{M2j}^{a}$ and ω_j^a are determined by using equation (5.3.12), and $\mathbf{U}_{N1j}^{g}$ and ω_j^g are determined by using equation (5.3.18). The solution given by equation (5.3.22) is valid only in the case where the nodal displacements are of small amplitude.

For large nodal displacements it is assumed that the nodal displacements take the form given by equation (4.3.21). They are composed of large nodal displacements $\mathbf{U}_l^0$ caused by rigid-body movement of members of the structure and small nodal displacements $\mathbf{U}_s^0$ caused by elastic deformation of members of the structure. $\mathbf{U}_l^0$ and $\mathbf{U}_s^0$ are given by equations (4.3.22) and (4.3.23) and the relationship given by equation (4.3.24) is valid. In the dynamical case, $\mathbf{U}_l^0$ and $\mathbf{U}_s^0$ are functions of time. By using $\mathbf{U}_l^0$ and $\mathbf{U}_s^0$, equation (4.3.26) takes the form

$$\mathbf{G}\mathbf{U}_s^0 + \mathbf{E}\mathbf{U}_l^0 + \mathbf{M}(\ddot{\mathbf{U}}_s^0 + \ddot{\mathbf{U}}_l^0) = \mathbf{Q} . \tag{5.3.23}$$

For free vibration this takes the form

$$\mathbf{G}\mathbf{U}_s^0 + \mathbf{E}\mathbf{U}_l^0 + \mathbf{M}(\ddot{\mathbf{U}}_s^0 + \ddot{\mathbf{U}}_l^0) = 0 . \tag{5.3.24}$$

Because of the relationship between $\mathbf{U}_l^0$ and $\mathbf{U}_s^0$ given by equation (4.3.4), **G** and **E** are functions of $\mathbf{U}_l^0$; the effect of $\mathbf{U}_s^0$ on the elements of these matrices is negligible and is ignored. Equation (5.3.24) can be divded into two sets of equations. One set takes the form

$$\mathbf{G}_{11}\mathbf{U}_s - (\mathbf{E}_{11}\mathbf{W}^T - \mathbf{E}_{12})\mathbf{U}_l + \mathbf{M}_{11}\ddot{\mathbf{U}}_s - \mathbf{M}_{11}\mathbf{W}^T\ddot{\mathbf{U}}_l = 0 \tag{5.3.25}$$

and the other set

$$\begin{aligned}&(\mathbf{G}_{21} - \mathbf{W}\mathbf{G}_{11})\mathbf{U}_s + (\mathbf{E}\mathbf{F}_{22} - \mathbf{W}\mathbf{E}_{12})\mathbf{U}_l - (\mathbf{E}_{21} - \mathbf{W}\mathbf{E}_{11})\mathbf{W}^T\mathbf{U}_l \\ &- \mathbf{W}\mathbf{M}_{11}\ddot{\mathbf{U}}_s + (\mathbf{M}_{22} + \mathbf{W}\mathbf{M}_{11}\mathbf{W}^T)\ddot{\mathbf{U}}_l = 0\ .\end{aligned} \tag{5.3.26}$$

In the cases where equation (4.3.24) is satisfied, the nature of matrix **G** implies that the contribution of the first term in equation (5.3.26) can be ignored and equation (5.3.26) then takes the form

$$\begin{aligned}&(\mathbf{E}\mathbf{F}_{22} - \mathbf{W}\mathbf{E}_{12})\mathbf{U}_l - (\mathbf{E}_{21} - \mathbf{W}\mathbf{E}_{11})\mathbf{W}^T\mathbf{U}_l \\ &- \mathbf{W}\mathbf{M}_s\ddot{\mathbf{U}}_s + (\mathbf{M}_{22} - \mathbf{W}\mathbf{M}_{11}\mathbf{W}^T)\ddot{\mathbf{U}}_l = 0\ .\end{aligned} \tag{5.3.27}$$

For moderately large nodal displacements, where the effect of the nodal displacements on matrix **G** and the contribution of the second term of equation (5.3.25) can be ignored, equation (5.3.25) takes the form

$$\begin{aligned}&\mathbf{D}_{11}\mathbf{U}_s + \mathbf{M}_{11}\ddot{\mathbf{U}}_s - \mathbf{M}_{11}\mathbf{W}^T\ddot{\mathbf{U}}_l = 0 \\ &\mathbf{D}_{11} = \mathbf{A}_{11}\mathbf{S}_{11}\mathbf{A}_{11}^T + \mathbf{A}_{12}\mathbf{S}_{22}\mathbf{A}_{12}^T\end{aligned}\ . \tag{5.3.28}$$

Here, it is also possible to distinguish between the cases of high and low frequencies, where the nodal displacements take the form of a harmonic function given by equation (5.1.4). By using equation (5.1.4), equation (5.3.27) takes the form

$$\begin{aligned}&(\mathbf{E}_{22} - \mathbf{W}\mathbf{E}_{12})\mathbf{U}^g_{lMj} - (\mathbf{E}_{21} - \mathbf{W}\mathbf{E}_{11})\mathbf{W}^T\mathbf{U}^g_{lMj} \\ &- (\omega^g_j)^2[(\mathbf{U}_{22} + \mathbf{W}\mathbf{M}_{11}\mathbf{W}^T)\mathbf{U}^g_{lMj} - \mathbf{W}\mathbf{M}_{11}\mathbf{U}^g_{sMj}] = 0\ .\end{aligned} \tag{5.3.29}$$

For high frequencies, where the third term of equation (5.3.29) is much larger than the other two terms,

$$\begin{aligned}&\mathbf{U}^g_{lMj} = \mathbf{L}\mathbf{U}^g_{sMj} \\ &\mathbf{L} = (\mathbf{M}_{22} + \mathbf{W}\mathbf{M}_{11}\mathbf{W}^T)^{-1}\mathbf{W}\mathbf{M}_{11}\ .\end{aligned} \tag{5.3.30}$$

By using equation (5.3.30), equation (5.3.28) takes the form

$$\mathbf{D}_{11}\mathbf{U}^g_{sMj} - (\omega^g_j)^2(\mathbf{M}_{11} - \mathbf{M}_{11}\mathbf{W}^T\mathbf{L})\mathbf{U}^g_{sMj} = 0\ . \tag{5.3.31}$$

Equation (5.3.31) is a self-adjoint equation, which can be used to determine $\mathbf{U}^{g}_{sMj}$ and ω^{g}_{j}. The number of frequencies and modes is equal to the number of elements of vector $\mathbf{U}^{g}_{sMj}$, i.e. r. The nodal displacements associated with the high frequencies $\mathbf{U}^{g}$ take the form

$$\mathbf{U}^{g} = \sum_{j}^{r} \begin{bmatrix} (\mathbf{I} - \mathbf{W}^{T}\mathbf{L})\ \mathbf{U}^{g}_{sMj} \\ \mathbf{L}\mathbf{U}^{g}_{sMj} \end{bmatrix} \sin(\omega^{g}_{j}t + \phi^{g}_{j}) \ . \tag{5.3.32}$$

For low frequencies of moderately large nodal displacements where the second and third terms of equation (5.3.28) can be ignored,

$$\mathbf{D}_{11}\mathbf{U}^{a}_{s} = 0 \ . \tag{5.3.33}$$

In this case,

$$\mathbf{U}^{a}_{s} = 0 \tag{5.3.34}$$

and equation (5.3.27) takes the form

$$(\mathbf{E}_{22} - \mathbf{W}\mathbf{E}_{12})\mathbf{U}^{a}_{l} - (\mathbf{E}_{21} - \mathbf{W}\mathbf{E}_{11})\mathbf{W}^{T}\mathbf{U}^{a}_{l} + (\mathbf{M}_{22} + \mathbf{W}\mathbf{M}_{11}\mathbf{W}^{T})\ddot{\mathbf{U}}^{a}_{l} = 0 \ . \tag{5.3.35}$$

Because the elements of matrix $\mathbf{E}$ are a function of $\mathbf{U}^{a}_{l}$, equation (5.3.35) indicates a non-linear vibration at low frequencies. The frequencies of vibration depend on the amplitude of the displacements. The analysis of non-linear vibration is beyond the scope of this book and the reader is referred to books on advanced analysis of vibration to analyse the response of the structure at low frequencies.

For large nodal displacements the dynamical response of the cabled structure is given by equations (5.3.25) and (5.3.26) and the approximation used for the analysis of moderately large nodal displacements is not valid and the solution given by equations (5.3.32), (5.3.34) and (5.3.35) is not correct. In this case, equation (5.3.25) is rearranged to take the form

$$\mathbf{G}_{11}\mathbf{U}_{s} + \mathbf{M}_{11}\ddot{\mathbf{U}}_{s} = (\mathbf{E}_{11}\mathbf{W}^{T} - \mathbf{E}_{12})\mathbf{U}_{l} + \mathbf{M}_{11}\mathbf{W}^{T}\ddot{\mathbf{U}}_{l} \ . \tag{5.3.36}$$

For large nodal displacements where condition (4.3.26) is satisfied and the elements of vector $\frac{1}{2}\mathbf{ASO}^{T}\mathbf{U}^{0}_{l}$ are much larger than the elements of vector $\mathbf{OSA}^{T}\mathbf{U}^{0}_{s}$, equation (5.3.36) takes the form

$$\mathbf{D}_{11}\mathbf{U}_{s} + \mathbf{M}_{11}\ddot{\mathbf{U}}_{s} = (\mathbf{E}_{11}\mathbf{W}^{T} - \mathbf{E}_{12})\mathbf{U}_{l} + \mathbf{M}_{11}\mathbf{W}^{T}\dot{\mathbf{U}}_{l} \ . \tag{5.3.37}$$

Equation (5.3.37) can be regarded as an equation of forced vibration; the force takes the form $(\mathbf{E}_{11}\mathbf{W}^{\mathrm{T}}\mathbf{E}_{12})\mathbf{U}_{\mathrm{l}} + \mathbf{M}_{11}\mathbf{W}^{\mathrm{T}}\ddot{\mathbf{U}}_{\mathrm{l}}$ of the following free-vibration case:

$$\mathbf{D}_{11}\mathbf{U}_{\mathrm{s}} + \mathbf{M}_{11}\ddot{\mathbf{U}}_{\mathrm{s}} = 0 \ . \tag{5.3.38}$$

Equation (5.3.38) can be used to determine the frequencies ω_{sj} and modes $\mathbf{U}_{\mathrm{sN}j}$ of free vibration. By using these modes and frequencies together with Duhamel's integral given by equation (5.2.11), $\mathbf{U}_{\mathrm{s}}$ can be formulated as a function of $\mathbf{U}_{\mathrm{l}}$ and $\ddot{\mathbf{U}}_{\mathrm{l}}$. By using the formulation, equation (5.3.27) takes the form of non-linear vibration of $\mathbf{U}_{\mathrm{l}}$.

5.4 AN UNDER-CONSTRAINED CABLED STRUCTURE: FORCED VIBRATION

The dynamical response of the under-constrained cabled structure takes the form given by equation (5.3.1). For small nodal displacements, equation (5.3.1) takes the form

$$\mathbf{DU} + \mathbf{VU} + \mathbf{M}\ddot{\mathbf{U}} = \mathbf{Q} \ . \tag{5.4.1}$$

By using the solution of free vibration given by equation (5.3.22), the modal matrix $\mathbf{U}_{\mathrm{N}}$ takes the form

$$\mathbf{U}_{\mathrm{N}} = \begin{bmatrix} -\mathbf{W}^{\mathrm{T}}\mathbf{U}_{\mathrm{N2}}^{\mathrm{a}} & \mathbf{U}_{\mathrm{N1}}^{\mathrm{g}} \\ \mathbf{U}_{\mathrm{N2}}^{\mathrm{a}} & \mathbf{N}\mathbf{U}_{\mathrm{N1}}^{\mathrm{g}} \end{bmatrix} . \tag{5.4.2}$$

Matrices $\mathbf{U}_{\mathrm{N1}}^{\mathrm{g}}$ and $\mathbf{U}_{\mathrm{N2}}^{\mathrm{a}}$ are formed by placing modes $\mathbf{U}_{\mathrm{N1}j}^{\mathrm{g}}$ and $\mathbf{U}_{\mathrm{N2}j}^{\mathrm{a}}$ one next to the other. Column j of matrix $\mathbf{U}_{\mathrm{N1}}^{\mathrm{g}}$ is the mode $\mathbf{U}_{\mathrm{N1}j}^{\mathrm{g}}$ and column j of matrix $\mathbf{U}_{\mathrm{N2}}^{\mathrm{a}}$ is the mode $\mathbf{U}_{\mathrm{N2}j}^{\mathrm{a}}$.

By using matrix $\mathbf{U}_{\mathrm{N}}$ it is possible to define the principal nodal displacements $\mathbf{U}_{\mathrm{P}}$ as given by equation (5.2.2) and (5.2.3). By using $\mathbf{U}_{\mathrm{P}}$, equation (5.4.1) takes the form given by equation (5.2.4). By using Duhamel's integral $\mathbf{U}_{\mathrm{P}j}$, it takes the form given by equation (5.2.11). Equation (5.2.11) indicates that in the case where $(\omega^{\mathrm{a}})^2 \ll (\omega^{\mathrm{g}})^2$ the principal nodal displacements associated with low frequencies are much larger than those associated with high frequencies. Only in the case where $Q_{\mathrm{P}j}^{\mathrm{a}}$ (associated with low frequencies ω^{a}) vanishes are these large displacements eliminated. In the case where mode j at low frequencies is $\mathbf{U}_{\mathrm{N}j}^{\mathrm{a}}$, $Q_{\mathrm{P}j}^{\mathrm{a}}$ (associated with low frequencies) takes the form

$$Q_{\mathrm{P}j}^{\mathrm{a}} = (\mathbf{U}_{\mathrm{N}j}^{\mathrm{a}})^{\mathrm{T}}\mathbf{Q} \ . \tag{5.4.3}$$

By using equation (5.3.18), equation (5.4.3) takes the form

$$Q^{a}_{Pj} = (\mathbf{U}^{a}_{N2j})^{T}(\mathbf{Q}_2 - \mathbf{W}\mathbf{Q}_1) \ . \tag{5.4.4}$$

$Q^{a}_{Pj} = 0$ where

$$\mathbf{Q}_2 = \mathbf{W}\mathbf{Q}_1 \ . \tag{5.4.5}$$

Condition (5.4.5) is the condition of the 'fitted load' given by equation (4.2.6). Equation (5.4.5) indicates that only for the 'fitted load' do large nodal displacements associated with low frequencies of vibration vanish.

In the case where the load causes moderately large nodal displacements, the response of the structure can be found by using equations (5.3.28) and (5.3.27) which take the form

$$\mathbf{D}_{11}\mathbf{U}_s + \mathbf{M}_{11}\ddot{\mathbf{U}}_s - \mathbf{M}_{11}\mathbf{W}^T\ddot{\mathbf{U}}_l = \mathbf{Q}_1 \tag{5.4.6}$$

and

$$\begin{aligned}&(\mathbf{E}_{22} - \mathbf{W}\mathbf{E}_{12})\mathbf{U}_l - (\mathbf{E}_{21} - \mathbf{W}\mathbf{E}_{11})\mathbf{W}^T\mathbf{U}_l \\ &- \mathbf{W}\mathbf{M}_{11}\ddot{\mathbf{U}}_s + (\mathbf{M}_{22} + \mathbf{W}\mathbf{M}_{11}\mathbf{W}^T)\ddot{\mathbf{U}}_l = \mathbf{Q}_2 - \mathbf{W}\mathbf{Q}_1 \ .\end{aligned} \tag{5.4.7}$$

Equation (5.4.6) can be rearranged to take the form

$$\mathbf{D}_{11}\mathbf{U}_s + \mathbf{M}_{11}\ddot{\mathbf{U}}_s = \mathbf{Q}_1 + \mathbf{M}_{11}\mathbf{W}^T\ddot{\mathbf{U}}_l \tag{5.4.8}$$

which can be considered a case of forced vibration of the system given by equation (5.3.38). By using equation (5.3.39), $\mathbf{U}_{sNj}$ and ω_{sj} can be determined. By using Duhamel's integral given by equation (5.2.11), $\mathbf{U}_s$ as a function of $\mathbf{U}_l$ and $\ddot{\mathbf{U}}_l$ can be formulated. By using the formulation, equation (5.4.7) takes the form of non-linear vibration of $\mathbf{U}_l$.

In the case where the nodal displacements are large, the response of the structure can be determined by using equations (5.4.7) and (5.3.37), which for dynamical loading takes the form

$$\mathbf{D}_{11}\mathbf{U}_s + \mathbf{M}_{11}\ddot{\mathbf{U}}_s = \mathbf{Q}_1 + (\mathbf{E}_{11}\mathbf{W}^T - \mathbf{E}_{12})\mathbf{U}_l + \mathbf{M}_{11}\mathbf{W}^T\ddot{\mathbf{U}}_l \ . \tag{5.4.9}$$

Here, also the same method used for moderately large nodal displacements can be followed. The only difference is that Duhamel's integral should include the second term on the right-hand side of equation (5.4.9).

Equations (5.4.6) and (5.4.9) show that the non-linear aspect of the dynamical behaviour of a cabled structure is associated with the deformation of the structure cased by the rigid-body movement of the members.

6

The design of a cabled structure

6.1 THE PRELIMINARY DESIGN AND CONFIGURATION OF UNDER-CONSTRAINED STRUCTURES

Under-constrained cabled structures are most attractive to the designer. The small number of elements makes them cost effective and the small number of joints saves on the workforce. Unfortunately they are the most difficult to design.

The various aspects which should be considered in the design of a cable net were discussed in section 1.1. It was shown that the cable net should consist of two families of cables with opposite curvatures. Examples of a cable net in the shape of a saddle, a cable net supported by a central mast and a cable net supported by an arch are shown in Figs 1.1.5, 1.1.6 and 1.1.7, respectively.

The preliminary design of a tensegric shell was discussed in section 1.2. Various tensegric nets are presented in Fig. 1.2.4 and the method of fitting these nets to tensegric shells of different shapes — a spherical shell, a cooling tower, a shallow spherical shell and a cylindrical shell — are shown in Figs 1.2.1, 1.2.2, 1.2.5 and 1.2.6, respectively. The preliminary design of the cabled structure determines its preliminary configuration, the location of the nodes and the position of the elements of the structure.

For example, the preliminary configuration of a simple tensegric shell composed of four bars and eight cables takes the form shown in Fig. 6.2.1.

6.2 THE PRESTRESSABLE CONFIGURATION

It was shown in section 3.1 that a cable net or a tensegric shell is suitable for practical purposes where it is prestressable. A structure of this type which does not satisfy this requirement is unacceptable and could not be constructed. It was shown in section 3.2 that the feasibility of prestressing the structure is associated with the rank of the geometrical matrix $\mathbf{A}$ of the structure given by equation (2.1.6). The geometrical matrix of the preliminary configuration of the structure is denoted by $\mathbf{D}_0$. For

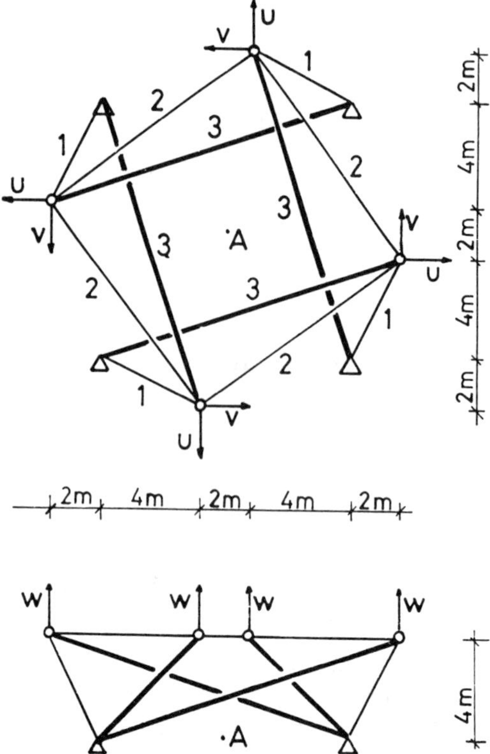

Fig. 6.2.1 — A preliminary configuration of a simple tensegric shell.

example, the geometrical matrix $\mathbf{D}_0$ of the preliminary configuration of the tensegric shell shown in Fig. 6.2.1, considering symmetrical prestressing, where identical forces are induced in members with identical numbers, takes the form

$$\mathbf{D}_0 = \begin{bmatrix} 0.33 & 1.40 & 0.91 \\ 0.67 & -0.20 & 0.30 \\ 0.67 & 0.00 & 0.30 \end{bmatrix} . \tag{6.2.1}$$

It was proved in section 3.2 that an under-constrained structure is prestressable and the structure is acceptable where the rank of matrix $\mathbf{D}_0$, denoted by r, is less than n, the number of the columns of that matrix:

$$r < n \,, \qquad \text{prestressable.} \tag{6.2.2}$$

For an under-constrained cable net or tensegric shell where equation (6.2.2) is satisfied, the preliminary configuration is prestressable and the preliminary configuration is identical with the prestressable configuration. In these cases the preliminary design of the structure is satisfactory and the prestressing forces, the effect of external forces and the dynamical response of the designed structure can be analysed by following Chapters 3, 4 and 5.

In most practical cases the preliminary configuration of an under-constrained structure does not satisfy equation (6.2.2). In these cases the preliminary configuration is not acceptable and should be altered to obtain a prestressable one.

For example, it is easy to realize that the preliminary configuration of the tensegric shell shown in Fig. 6.2.1 is not prestressable. The rank of matrix $\mathbf{D}_0$ given by equation (6.2.1) is 3; n in this case is also equal to 3 and thus equation (6.2.2) is not satisfied. In these cases the prestressable configuration is different from the preliminary configuration. A prestressable configuration may be achieved from the preliminary configuration by considering some nodal displacements. The nodal displacements affect the geometrical matrix $\mathbf{D}_0$. They change the geometrical matrix from $\mathbf{D}_0$ to $\mathbf{A}$, in which $\mathbf{A}$ is the geometrical matrix of a prestressable configuration. The change in the configuration of member i due to the nodal displacements is shown in Fig. 3.4.1. $\cos\alpha_{ix}$ changes to $\cos\alpha'_{ix}$ owing to the nodal displacements as given by equation (3.4.28). In this case the geometrical matrix of the prestressable structure can be seen as a linear combination of the geometrical matrix $\mathbf{D}_0$ considering the preliminary configuration and matrix $\mathbf{O}$. The elements of matrix $\mathbf{O}$ are functions of the nodal coordinates of the preliminary configuration and the nodal displacements given by the last three terms of equation (3.4.28). In this case, matrix $\mathbf{A}$ takes the form

$$\mathbf{A} = \mathbf{D}_0 + \mathbf{O} \ . \tag{6.2.3}$$

For the tensegric shell shown in Fig. 6.2.1, considering the case of symmetrical nodal displacements shown there, matrix $\mathbf{O}$ takes the form

$$\mathbf{O} = 10^{-2}\begin{bmatrix} 7.40u - 1.85v - 1.85w & 0.40u + 2.80v & 1.37u - 4.10v - 4.10w \\ -3.70u - 9.26v - 7.40w & 2.80u + 19.60v & -2.06u + 6.17v - 1.37w \\ -3.70u - 7.40v - 9.25w & 0.0 & -2.06u - 1.37v + 6.17w \end{bmatrix} . \tag{6.2.4}$$

u, v and w are measured in metres. For small nodal displacements, the magnitudes of the elements of matrix $\mathbf{O}$ are much smaller than the magnitudes of the elements of matrix $\mathbf{D}_0$. Equation (6.2.3) implies that

$$\operatorname{rank}(\mathbf{D}_0 + \mathbf{O}) < n \ . \tag{6.2.5}$$

Equation (6.2.5) is satisfied where

$$\operatorname{rank}(\mathbf{D}_0 + \mathbf{O}) = n - 1 \ . \tag{6.2.6}$$

Equation (6.2.6) is satisfied where all possible determinates of $n \times n$ formed appropriately from matrix $\mathbf{A}$ are equal to zero. In the general case there are p independent determinants:

$$p = m - n + 1 \ . \tag{6.2.7}$$

Equation (6.2.7) takes the form:

$$\text{def}(\mathbf{A}^{nk} + \mathbf{O}^{nk}) = 0 \ , \qquad k = 1,p \ . \tag{6.2.8}$$

$\mathbf{A}^{nk}$ and $\mathbf{O}^{nk}$ are determinate k of $n \times n$ formed from matrix **A** and **O** appropriately. Because of the relationship between the magnitudes of the elements of determinates $\mathbf{A}^{nk}$ and $\mathbf{O}^{ak}$, equation (6.2.8) takes the form

$$\sum_i \sum_j \mathbf{A}_{ij}^{nk} \mathbf{CA}_{ij}^{nk} + \sum_i \sum_j \mathbf{O}_{ij}^{nk} \mathbf{CA}_{ij}^{nk} \ , \qquad k = 1,p \ . \tag{6.2.9}$$

$\mathbf{A}_{ij}^{nk}$ is the element ij of matrix $\mathbf{A}^{nk}$, $\mathbf{CA}_{ij}^{nk}$ is the cofactor of elements ij of the same matrix and $\mathbf{O}_{ij}^{nk}$ is element ij of matrix $\mathbf{O}^{nk}$. Equation (6.2.9) yields p linear equations of the nodal displacements.

For example, for the tensegric shell shown in Fig. 6.2.1 where equations (6.2.1) and (6.2.5) are considered, equation (6.2.9) yields one linear equation only:

$$0.102 + 0.025u + 0.150v - 0.026w = 0 \ . \tag{6.2.10}$$

Equation (6.2.10) indicates the required relationship between the various nodal displacements in order to achieve a prestressable configuration from the preliminary configuration. It can be seen that the requirement of a prestressable configuration is not enough to determine the nodal displacements uniquely. In this example there are three degrees of freedom to the structure (the displacements u, v, w). A prestressable configuration implies only one relationship given by equation (6.2.10) between them. This result is typical of an under-constrained structure and it indicates the freedom of the designer in obtaining the prestressable configuration. The designer can add additional constraints to the nodal displacements with the change in the structure from its preliminary configuration to its prestressable configuration.

6.3 CONSTRAINTS ON THE DESIGNED CONFIGURATION

In the process of obtaining the prestressable configuration the designer may wish to change the length of some members only and to keep certain members intact. In the case where member i shown in Fig. 3.4.1 is designed to maintain its length l_i^0, while the structure configuration is altered from a preliminary one to the prestressable one:

$$[l_{ix} + (u_4^i - u_1^i)]^2 + [l_{iy}(u_5^i - u_2^i)]^2 + [l_{iz} + (u_6^i - u_5^i)]^2 = (l_i^0)^2 \tag{6.3.1}$$

$$l_{ix} = l_i^0 \cos \alpha_{ix} \ .$$

For small nodal displacements, equation (6.3.1) takes the form

$$l_{ix}(u_4^i - u_1^i) + l_{iy}(u_5^i - u_2^i) + l_{iz}(u_6^i - u_5^i) = 0 \ . \tag{6.3.2}$$

Equation (6.3.2) is a linear equation of the nodal displacements associated with the change in the configuration of the structure.

For example, if members 2 of the tensegric shell shown in Fig. 6.2.1 are designed to maintain their length, equation (6.3.2) takes the form

$$7u - v = 0 \ . \tag{6.3.3}$$

The designer may wish to maintain other dimensions. For example, for the

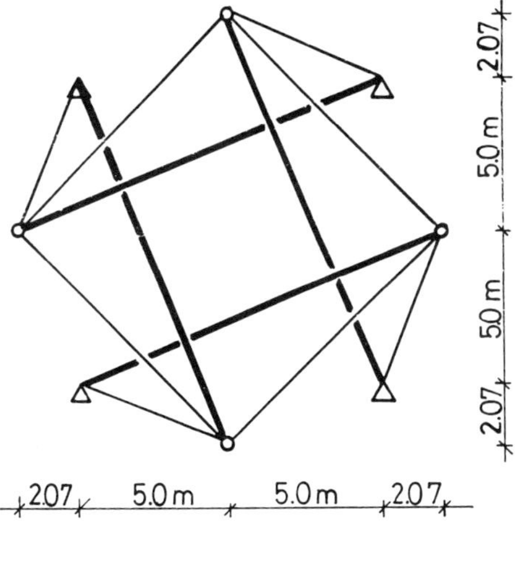

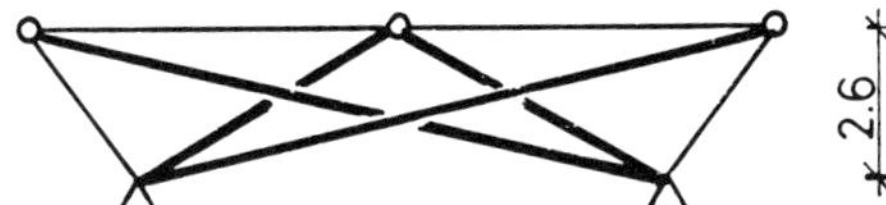

Fig. 6.3.1 — The prestressed configuration.

tensegric shell shown in Fig. 6.2.1 the designer may maintain the distance between point A and the nodes. In this case this condition implies that

$$(7-u)^2+(1-\upsilon)^2-(4-w)^2=66 \ . \tag{6.3.4}$$

For small nodal displacements, equation (6.3.4) takes the form

$$7u-\upsilon+4w=0 \ . \tag{6.3.5}$$

Another constraint that can be considered is to minimize the nodal displacement *DI*, from the preliminary configuration. In this example it takes the form

$$DI=u^2+\upsilon^2+w^2 \ . \tag{6.3.6}$$

In the case where the designer wishes to maintain the length of members 2 and to minimize the nodal displacement *DI* equations (6.2.9) and (6.3.3), equation (6.3.6) implies

$$\frac{\partial DI}{\partial u}=2u+14\upsilon+82.6w=0 \ . \tag{6.3.7}$$

In this case the three equations (6.2.9), (6.3.3) and (6.3.7) can be used to define the presentable configuration considering these constraints.

In cases where the transition from the preliminary configuration to the prestressed configuration is associated with large nodal displacements the linear approximate analysis discussed above is not accurate. In these cases this method should be followed step by step. The analysed prestressed configuration should be considered as a preliminary configuration for the next step of analysis. The analysis should be followed until the change in the prestressed configuration from the preliminary one is within the required level of accuracy.

The prestressed configuration of the preliminary configuration shown in Fig. 6.2.1 takes the form shown in Fig. 6.3.1 where elements 1 and 2 are considered to maintain their length. In this case, matrix **A** takes the form

$$\mathbf{A} = \begin{bmatrix} 0.35 & 1.41 & 0.90 \\ 0.82 & 0.00 & 0.38 \\ 0.43 & 0.00 & 0.20 \end{bmatrix} . \qquad (6.3.8)$$

The analysis of the forces induced by prestressing of the shell is carried out by following the method presented in Chapter 3. The structure is acceptable only where tension is induced in the cables by prestressing.

The analysis of the structure considering statical and dynamical loading is given in Chapters 4 and 5.

Appendix I
The theory of a single cable

A cable cannot sustain compression and it does not have flexural rigidity; it can sustain tensile internal forces only. When loaded it takes the typical shape shown in Fig. I.1. The configuration of the cable is presented by function w:

$$w = w(x) \ . \tag{I.1}$$

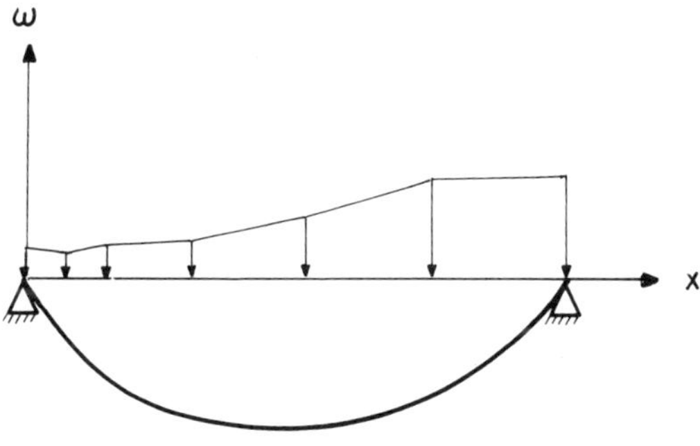

Fig. I.1 — A cable.

In the case where the cable is loaded by vertical load only, a differential element of the cable takes the shape shown in Fig. I.2. T_1 and T_2 are the tensile tensional forces in the cable at points 1 and 2. H and $H + (\mathrm{d}H/\mathrm{d}x)\,\mathrm{d}x$ are the horizontal components of the tensile force at these points. The equilibrium condition of statics of the horizontal forces takes the form

$$H + \frac{\mathrm{d}H}{\mathrm{d}x}\mathrm{d}x - H = 0 \ . \tag{I.2}$$

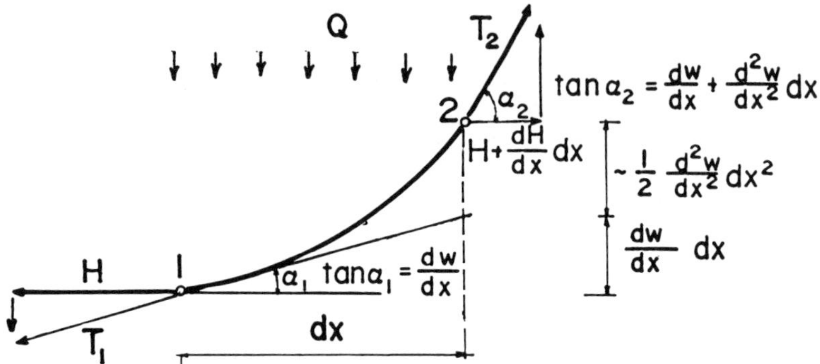

Fig. I.2 — A differential element of the cable.

Equation (I.2) indicates that

$$\frac{\mathrm{d}H}{\mathrm{d}x} = 0 \tag{I.3}$$

and so

$$H = H_0 \tag{I.4}$$

is a constant. Equation (I.4) implies that, in the case where the cable is loaded by vertical load only, the horizontal component of the internal tensile force is constant. By using equation (I.4) the equilibrium condition of statics of the vertical forces is

$$H_0\left(\frac{\mathrm{d}w}{\mathrm{d}x} + \frac{\mathrm{d}^2w}{\mathrm{d}x^2}\mathrm{d}x\right)\mathrm{d}x - H_0\frac{\mathrm{d}w}{\mathrm{d}x} = Q\,\mathrm{d}x \tag{I.5}$$

which takes the form

$$H_0\frac{\mathrm{d}^2w}{\mathrm{d}x^2} = Q\ . \tag{I.6}$$

Equation (I.6) is the **cable equation**. The equilibrium condition of statics of the moments about point 2 takes the form

$$H_0\left(\frac{\mathrm{d}w}{\mathrm{d}x}\mathrm{d}x + \frac{1}{2}\frac{\mathrm{d}^2w}{\mathrm{d}x^2}\mathrm{d}x^2\right) - H_0\frac{\mathrm{d}w}{\mathrm{d}x}\mathrm{d}x = Q\frac{\mathrm{d}x^2}{2}\ . \tag{I.7}$$

It can be seen that the equation is similar to equation (I.5). The equilibrium condition of statics of the moments takes the same form as of the vertical forces.

The internal tension in the cable takes the form

$$T = H_0\sqrt{1 + \left(\frac{\mathrm{d}w}{\mathrm{d}x}\right)^2}\ . \tag{I.8}$$

Equation (I.6) indicates that the cable configuration is dictated by the nature of the load. Where the load changes, the configuration of the cable changes. It is easy to determine the configuration of the cable by using the bending theory of beams. The relationship between the bending moments M_b and the load of a beam takes the form

$$\frac{d^2M_b}{dx^2} = Q \ . \tag{I.9}$$

By comparing equation (I.9) with equation (I.6) it can be seen that w of a cable is analogous to M of a beam. The bending moment diagram of a beam indicates the configuration of the cable. For example, the bending moment diagram of a simply supported beam loaded by a uniformly distributed load has the typical parabolic shape shown in Fig. I.3(a). Equations (I.9) and (I.6) indicate that the configuration

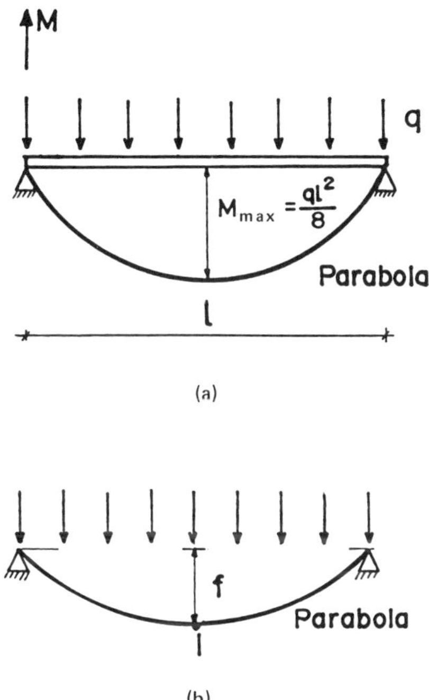

Fig. I.3 — (a) The bending moment diagram, and (b) the configuration of the cable.

of a cable loaded by a uniformly distributed load takes the parabolic shape too. In the case of a point load acting at the centre of the span, the cable takes the form shown in Fig. I.4(a). The case of two identical point loads at a third of the span is shown in Fig. I.4(b). Figs I.3 and I.4 show how the configuration of the cable changes when the load changes. These large changes in the cable configuration indicate that a cable does not have a **fixed geometry**.

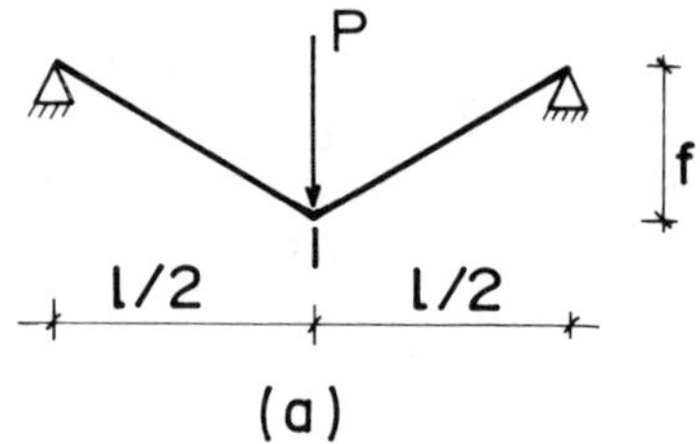

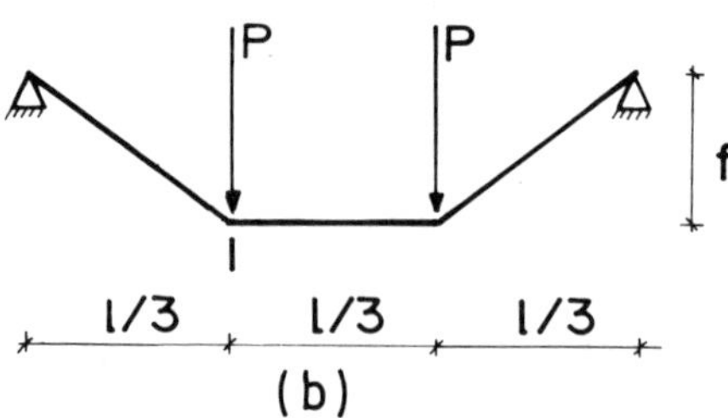

Fig. I.4 — Configuration of the cable in various loading cases.

The horizontal component of the forces applied to the supports is H_0. By considering the equilibrium conditions of statics the reaction at the supports can be determined as a function of P. H_0 can be found by taking the moments about any point on the cable to be equal to zero. For example, H_0 of the cable shown in Fig. I.3, where the moment about point 1 is considered, takes the form

$$H_0 = \frac{ql^2}{8f} . \tag{I.10}$$

In the case shown in Fig. I.4(a) (where the moment about point 1 is considered), it takes the form

$$H_0 = \frac{pl}{4f} \tag{I.11}$$

and, in the case shown in Fig. I.4(b) (where the moment about point 1 is considered), it takes the form

$$H_0 = \frac{pl}{3f} . \tag{I.12}$$

In every case the internal tension in the cable can be determined by using equation (I.8).

The stability of the cable is studied by investigating the change in its energy associated with a small distortion of its configuration. For example, the distortion of the cable shown in Fig. I.4(b) takes the form shown in Fig. I.5. For small nodal

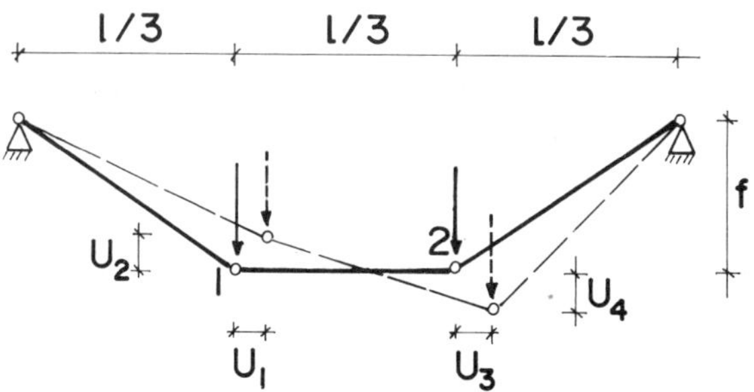

Fig. I.5 — Small distortion of the configuration of the cable.

displacements and where there is no change in the length of the cables,

$$\left(\frac{l}{3}+U_1\right)^2+(f-U_2)^2=\left(\frac{l}{3}\right)^2+f^2$$

$$\left(\frac{l}{3}\quad U_3\right)^2+(f+U_4)^2-\left(\frac{l}{3}\right)^2+f^2 \qquad \text{(I.13)}$$

$$\left(\frac{l}{3}+U_3-U_1\right)^2+(U_4-U_2)^2=\left(\frac{l}{3}\right)^2 .$$

By using the first two equations of equation (I.13),

$$U_1 \approx \frac{3f}{l}U_2$$

$$U_3 \approx \frac{3f}{l}U_4 . \qquad \text{(I.14)}$$

By using equation (I.14), the third equation of equation (I.13) takes the form

$$\left(\frac{l}{3}+\frac{3f}{l}(U_4-U_2)\right)^2+(U_4-U_2)^2=\left(\frac{l}{3}\right)^2 . \qquad \text{(I.15)}$$

In the case where it is assumed that

$$U_2 = U_4 + C_0(U_4)^2 \qquad \text{(I.16)}$$

(in which C_0 is a constant to be determined), equation (I.15) takes the form given by (orders higher than the second order of U_4 are ignored)

$$-2fC_0(U_4)^2+4(U_4)^2=0\ . \tag{I.17}$$

By using equation (I.17), C_0 takes the form

$$C_0=\frac{2}{f}\ . \tag{I.18}$$

Because no energy is associated with the rotation of the internal forces, the change E_a in the energy associated with the distortion is

$$E_a=PV_4-PV_2\ . \tag{I.19}$$

By using equations (I.16) and (I.18), equation (I.19) takes the form

$$E_a=-\frac{2P}{f}(U_4)^2\ . \tag{I.20}$$

An energy is required for the change in the configuration of the cable. This implies that this cable is *stable*. In every case where its configuration is distorted, it will return to its initial configuration. The cable is always stable and its stability is not associated with its rigidity as in the case of a simple strut.

The result is typical for all cables. A cable is always stable. The cable does not have a fixed configuration but every configuration caused by a specific load is a stable configuration.

The fact that cables are always stable makes them very attractive for bridging large spans. One of the most remarkable achievements in the structural engineering of our time is the use of cables to bridge large spans. Notable examples are suspension bridges.

Suggestions for further reading

Fuller, R. B., and Marks, R. (1973) *The dymasion world of Buckminster Fuller.* Anchor, New York.

Høllmann, H. (1965) *A study in the theory of suspension structures.* Akademish Forlag, Denmark.

Irvine, H. M. (1981) *Cable structures.* MIT Press, Cambridge, Massachusetts.

Kenner, H. (1976) *Geodesic maths and how to use it.* University of California Press, Berkeley and Los Angeles, California.

Krishna, P. (1978) *Cable-suspended roofs.* McGraw-Hill, New York.

Otto, F. (1954) *Das Hangende Dach.* Bauwelt, Berlin.

Otto, F. (1973) *Tensile structures.* MIT Press, Cambridge, Massachusetts.

Pugh, A. (1976) *An introduction to tensegrity*, University of California Press, Berkeley and Los Angeles, California.

Rabinovič, I. M. (1966) *Hängedächer*, Bauvelag, Wiesbaden-Berlin.

Szabó, J., and Kollár, L. (1984) *Structural design of cable-suspended roofs.* Ellis Horwood, Chichester, West Sussex.

Index

angular frequency, 98

bending-moment diagram, 123

cable equation, 122
cable nets
  advantages, 11
  comparison with tensegric shells, 15
  disadvantages, 11
  feasible, requirements of, 33
  fully constrained, 34, 36–37
  supported by a mast, 11, 12, 13, 115
  supported by an arch, 11, 13, 115
  under-constrained, 37
cabled structures, definition, 15
configuration
  of cable nets, 9–12
    effect of wind uplift, 9–10
    prestressable, 115–118
  constraints on, 118–120
  increasing the self-weight of the roof, 10
  large change in, 11
  small distortion of, 9, 125
  of tensegric shells, 12–19
    prestressable, 115–118
  in various loading cases, 124
connector in tensegric shells, 15, 19

deformation matrix, 22–25
  relationship between equilibrium matrix and, 25–27
deformed truss, 23
design of cabled structure, 115–120
determinate structure, 28, 29
diagonal matrix, 28
Duhamel's integral, 104, 105, 113, 114
dynamical behaviour, 97–114

eigenfunctions, 98
eigenvalues, 98
equilibrium matrix, 20–22
  relationship between deformation matrix and, 25–27
external load, effect of, 72–96
  fitted, 73–83
  non-fitted, 83–96

fitted load, 73–83
  cable net loaded by, 75, 77, 78
  change in internal forces of cable net, 79, 80
  definition, 75
  nodal displacements of cable net due to, 83, 84
  tensegric shell loaded by, 76
fixed geometry, 31, 123
forced vibration
  of fully constrained cabled structure, 103–105
  of under-constrained cabled structure, 113–114
free nodes, 21
free vibration
  of fully constrained cabled structure, 97–103
  of under-constrained cabled structure, 106–113
fully constrained cabled structures, 58
  cable nets, 34, 36–37
  effect of external load, 72–73
  forced vibration, 103–105
  free vibration, 97–103
  tansegric shells, 34, 35

Gaussian elimination, 38–40, 42, 48, 51–53
generalized stiffness matrix, 29
geometrical non-linear effect, 62

Hooke's law, 27
hyperbolic paraboloid cable net, 49, 50

indeterminate structure, 29

kinematic degree of freedom, 22

mechanism, 30, 31

nodal displacements
  in cable net due to fitted load, 83
  of cabled structure with fitted load, 82
  in deformation matrix, 22–24
  due to rigid-body movement, 83, 88–92
  linear combination, 64
  in pre-stressing, 58–62, 68, 69
  principal, 103
  of under-constrained cabled structures, 62–63

non-fitted load, 83–96

orthogonality of modes of vibration, 99

phase angle, 98
prestressable configuration, 115–118, 119
  constraints on, 118–120
prestressing
  of cable net, 10, 54, 55, 56, 69, 71
  deformation due to, 57–71
  feasibility of, 33–53
  forces induced by, 53–57
  importance of, 33
  of tensegric shell, 14, 54, 55, 68
  tension induced by, 56
principal nodal displacements, 103
principle of virtual work, 25

reticulated structures, analysis of, 28–32
rigid-body movement, 83, 88–92, 110

saddle roofs, 11, 12, 46, 47, 49, 115
self-adjoint equation, 98
single cable theory, 9, 121–126
singular matrix, 31, 32
spherical tensegric shell, 14, 16, 115
  shallow, 14, 18, 115
square cable nets, 14
square matrix, 27
stiffness matrix, 27–28

Taylor's series, 94
tensegric nets, 14
tensegric shell
  advantages, 14–15
  comparison with cable nets, 15
  cylindrical, 14, 18, 115
  definition, 12
  disadvantages, 15
  feasible, requirements of, 33
  fully constrained, 34, 35
  in shape of cooling tower, 14, 15, 115
triangular cable nets, 14
two-cable-system, 10, 11
two-dimensional truss, 21

under-constrained cabled structures, 37, 52, 53, 55, 59, 61
  analysis of, 87–96
    by symmetry, 82
  cable net, 37
  fitted load, 73–83
  free vibration, 106–113
  nodal displacements, 63–64
  non-fitted load, 83–96
  preliminary design and configuration, 115
  tensegric shells, 34, 35

vibration
  forced
    of fully constrained cabled structure, 103–105
    of under-constrained cabled structure, 113–114
  free
    of fully constrained cabled structure, 97–103
    of under-constrained cabled structure, 106–113